John Cooper Forster

The surgical diseases of children

John Cooper Forster

The surgical diseases of children

ISBN/EAN: 9783337215293

Printed in Europe, USA, Canada, Australia, Japan

Cover: Foto ©berggeist007 / pixelio.de

More available books at **www.hansebooks.com**

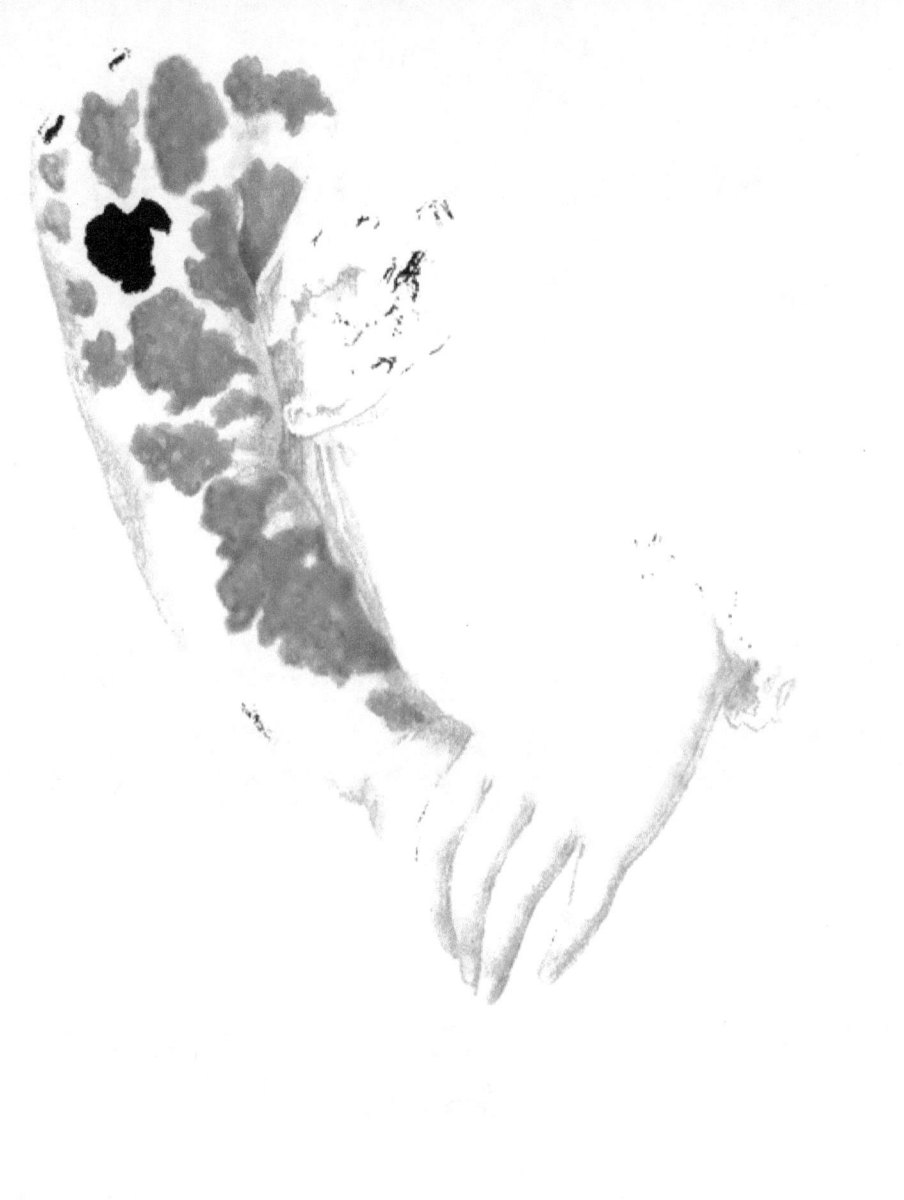

THE SURGICAL DISEASES

OF

CHILDREN.

BY

J. COOPER FORSTER,

FELLOW OF THE ROYAL COLLEGE OF SURGEONS; BACHELOR OF MEDICINE
OF THE UNIVERSITY OF LONDON;
ASSISTANT-SURGEON TO, AND LECTURER ON ANATOMY AT, GUY'S
HOSPITAL; AND
SURGEON TO THE ROYAL INFIRMARY FOR CHILDREN, ETC.

LONDON:
JOHN W. PARKER AND SON, WEST STRAND.
1860.

LONDON:
SAVILL AND EDWARDS, PRINTERS, CHANDOS STREET,
COVENT GARDEN.

TO

JOHN FORSTER, ESQ., F.L.S.,

This Volume is Dedicated

BY A GRATEFUL AND AFFECTIONATE SON.

PREFACE.

I HAVE long designed to lay before the Profession some remarks upon those affections of children—that is, of patients under puberty—the care of which comes within the province of the surgeon. The absence of any work in the English language devoted to this subject seems to render it not inappropriate for those who, like myself, have been largely engaged in the treatment of this class of diseases, to make public the results of their experience. Into the surgical wards of Guy's Hospital more than three hundred children are annually admitted; and at the Royal Infirmary for Children a very large number of similar patients have for many years come under my care. To treat these cases practically is one thing; to write a book upon the subject is quite another. I must, therefore, beg the reader's indulgence for any deficiencies or errors which may have found their way into this volume. One thing, however, I may promise—the entire matter of it is original. The work contains the results of my own experience. With one exception, all the illustrations have been taken from drawings made under my own superintendence, or that of my colleagues, and never before published. Moreover, almost every case referred to has passed under my own eye. Accordingly, the reader must not anticipate a complete systematic treatise upon the surgical diseases of childhood. Where my own observation does not afford any data, I have been silent; and if references to the labours of others are rarely made, this should be ascribed rather to the plan and scope of the work not admitting of them, than to a disregard of what has been accomplished.

I cannot, however, omit to express my thanks to Messrs. Cock and Hilton, my kind and valued instructors, and to my other colleagues at Guy's Hospital. Many of the cases adduced in illustration of the principles maintained have been taken from the wards over which they so ably preside.

If I had attempted fully to discuss *all* the surgical diseases to which children, in common with adults, are liable, this volume would necessarily have expanded into a complete treatise on surgery, and must have contained much that would have been a mere repetition of familiar knowledge. I have therefore dwelt but slightly upon some affections which do not present, in the case of children, any differences, either in character or treatment, from the same diseases in the adult. Among these are injuries of the head, hernia, &c. Brevity has been a constant object with me; feeling that, probably, some of my readers might be amongst those whom I have the pleasure to know as students of Guy's Hospital; and being well aware that they have so many claims upon their time as to justify them in expecting that all necessary information should be put before them in the most concise form.

Diseases of the eye have been omitted altogether, since they seem to be, by common consent, transferred to the care of those who especially devote themselves to Ophthalmic Surgery.

To my friend and colleague, Dr. Willshire, I am greatly indebted for very valuable assistance rendered in many ways.

J. COOPER FORSTER.

WELLINGTON STREET,
   LONDON BRIDGE, 1860.

CONTENTS.

CHAPTER I.

	PAGE
INTRODUCTORY REMARKS.	
Anæsthetics	1
Nursing	4
DISEASES AND INJURIES OF THE HEAD.	
Concussion—Fracture—Compression	5

CHAPTER II.

DISEASES AND INJURIES OF THE FACE.	
Wounds of the Face	11
Diseases and Injuries of the Nose	13
Foreign Bodies in the Nose	13
Epistaxis	16
Chronic Thickening of the Schneiderian Membrane	18
Polypus of the Nose	19
Strumous Ulceration of the Face	21
Cancrum Oris	23

CHAPTER III.

DISEASES OF THE MOUTH AND TONSILS.	
Hare Lip	29
Tongue-tie	36
Ranula	37
Cleft Palate	39
Diseases of the Gum	40
Tumour of the Antrum	40
Bony Tumour of the Upper or Lower Jaw	41
Strumous Disease of the Hard Palate	42
Disease of the Tonsils	43

CHAPTER IV.

AFFECTIONS OF THE LARYNX AND TRACHEA.

Accidents to the Larynx 50
Foreign Bodies in the Air Passages 55
The Operation of Tracheotomy 66

CHAPTER V.

AFFECTIONS OF THE PHARYNX AND ÆSOPHAGUS.

Diseases and Injuries of the Pharynx and Æsophagus 76
Foreign Bodies in the Pharynx, &c. 78

CHAPTER VI.

DISEASES OF THE RECTUM.

Hæmorrhoids or Piles 89
Prolapsus Ani 89
Polypus of the Rectum 93
Fistula in Ano 94

CHAPTER VII.

AFFECTIONS OF THE TRUNK.

Diseases of the Neck 99
Cysts in the Neck 103
Torticollis, or Wry-neck 104
Diseases of the Umbilicus 105
Growth from the Umbilicus 106
Fæcal Fistula at the Umbilicus 107
Disease of the Vertebræ—Angular Curvature . . . 111

CHAPTER VIII.

DISEASES OF THE URINARY AND GENERATIVE ORGANS IN THE FEMALE.

Discharges from the Vagina in Children 123
Strumous Ulceration of the Vagina 127
Noma 128
Closure of the Orifice of the Vagina 129
Villous Growth from the Bladder 130
Calculus in the Female 132

CHAPTER IX.

DISEASES OF THE URINARY AND GENERATIVE ORGANS IN THE MALE.

Ruptured Urethra	136
Calculus in the Urethra	139
Extravasation of Urine	145
Incontinence of Urine	155
Polypus of the Bladder	160

CHAPTER X.

CALCULUS IN THE BLADDER 163

CHAPTER XI.

DISEASES OF THE TESTIS.

Diseases of the Tunica Vaginalis and Cord	183
Diseases of the Testicle Proper	188
Malformations of the Penis	191
Gonorrhœa	196
Strumous Ulceration of the Penis	196
Onanism	197

CHAPTER XII.

HERNIA.

Umbilical Hernia	199
Inguinal Hernia	201

CHAPTER XIII.

NÆVUS 206

CHAPTER XIV.

TREATMENT OF NÆVUS 207

CHAPTER XV.

INJURIES AND DISEASES OF BONES.

Fractures	250
Dislocations	252
Diseases of Bone	253

CHAPTER XVI.

INJURIES AND DISEASES OF THE JOINTS.
    Injuries to the Joints 258
    Diseases of the Joints 259
    Contracted Joints 282

CHAPTER XVII.

DISEASES OF THE SKIN 286
INFANTILE SYPHILIS 290
DISEASES OF THE NAILS 294
WARTS AND CORNS 295

CHAPTER XVIII.

DISEASES OF THE EAR.
    Foreign Bodies in the Meatus 296
    Discharges from the Ear 298
    General Remarks 306

CHAPTER XIX.

CONGENITAL DEFORMITIES AND MALFORMATIONS.
    Encephalocele 308
    Spina Bifida 311
    Occlusion of the Mouth 315
    Imperforate Rectum 315
    Congenital Fractures and Dislocations 321
    Excess or Deficiency of Extremities 322
    Webbed Fingers 323
    Club Foot 324
    Hermaphroditism 325
    Defect of the Abdominal Parietes 326

CHAPTER XX.

TETANUS 328
SCALDS AND BURNS 334

LIST OF COLOURED PLATES.

1. Frontispiece—Extensive Nævus of the Arm.
2. Cancrum Oris . . . . *to face p.* 26
3. Bean lodged in the Right Bronchus 66
4. Vascular Growth from the Umbilicus 106
5. Nævus on the Forehead of a Boy; and Dissected Structure of Nævus 210
6. Nævus of the Lower Extremity, in parts undergoing spontaneous cure 213
7. Extensive Nævus beneath the Ear, affecting the Lobe; and small Nævus beneath the Eye 214
8. Nævus on the Lip; and on the Vulva 219
9. Subcutaneous Nævus, the surface ulcerating; and Hairy Nævus around the Eye 249
10. Spina Bifida 311

THE SURGICAL DISEASES OF CHILDREN.

CHAPTER I.

INTRODUCTORY REMARKS.

I. ANÆSTHETICS.

BEFORE proceeding to the special subject of this book, I may premise a few observations on Chloroform and Freezing Mixtures, especially since I shall have continually to allude to the use of the former. In fact, the surgery of children, particularly as regards its manipulative and operative branches, has made rapid strides since the introduction of means by which pain may be prevented. Any one who remembers the operations conducted years ago, before chloroform was known, cannot fail to be impressed with the great facilities which that agent gives to the operator, and its immense benefit to the patient. Indeed it appears to me that under present circumstances we are never justified in inflicting any severe pain upon children. Chloroform is so safe in the case of the young child, that I never perform even the slightest operation without using it, with the one exception of excising the tonsils. In this case I think it advisable not to incur the risk of bleeding into the larynx, and fortunately the sensibility of those organs is very small. I believe that as children advance in years the danger of chloroform increases, but among the thousands of cases in which it has been employed in early life, I have heard of very few fatal results from its direct effects. As children advance towards puberty, freezing mixtures may be had recourse to in suitable cases; but in the very young I never employ them. The pain connected with their application seems to be hardly less than that of the operation they are designed to mitigate.

Having expressed this general opinion respecting the use of chloroform, I need scarcely mention in detail the various cases in which I should recommend its exhibition. It is, of course, not only in using the knife that I have recourse to its soothing influence, but also in putting up fractures, and even in examining them, or diseased joints, or in any other painful manipulation for diagnostic purposes. Chloroform should always be given on an empty stomach, to avoid sickness. The mode in which I administer it is, by pouring a few drops on four or five folds of lint covered with oiled silk; this is then held loosely over the little patient's mouth and nose, taking care not to touch the nose, as chloroform applied to the skin produces a very unpleasant blister. The children seem to enjoy it, and in the course of a minute or two, if the chloroform is good, they begin to yield to its anæsthetic influence. Sickness is rarely produced in them, and when the lint is withdrawn they seem to lie in a comfortable sleep. The sensitiveness of the conjunctiva is the test on which I rely.

From the foregoing remarks it will easily be understood that, so far as children are concerned, I have no belief in the injurious after effects which have been sometimes ascribed to chloroform. If any mischief results, I believe it is probably due to some error in the mode of its administration. Children are so readily susceptible of the influence of chloroform, that a little excess produces a marked effect upon them, but they also quickly rally on its withdrawal.

As regards the use of freezing mixtures, I believe they are quite inapplicable to very young children; still, even in them the use of ice, to which the child soon becomes accustomed, benumbs the sensibility of the parts, and allows them to be handled, or even incised, without a great amount of pain. It is in children verging on puberty that this agent is most advantageously employed, in the opening of abscesses, and other cases in which it is desirable to abolish the sensibility of the skin.

The best mode of applying cold is to mix thoroughly together equal parts of ice, broken into very small pieces, and of salt, and place them in a piece of very thin caoutchouc sheeting. This should be gently drawn to and fro over the part for two or three minutes, until the skin begins to assume a whitish appearance. In this state, if the skin is incised, it cuts like a raw turnip, and without the least sensation to the patient. Some little uneasiness attends the first application of the freezing mixture, and also its withdrawal. The latter may be somewhat relieved by making the withdrawal gradual. To avoid sloughing, care should be taken that the application be not continued too long.

II. NURSING.

When we remember the innumerable works which issue from the press under the titles of *Hints to Mothers*, &c., I feel some hesitation in saying anything upon this important subject; but the surgeon alone is able to appreciate to its full extent how much depends, especially for the success of operations, upon the careful and kindly attention of a loving nurse. A nurse to be fit to undertake the management of children, must almost be born a nurse, so much seems to depend upon an instinctive tact and insight into their little ways. Moreover, the very young cannot reward their attendant with gratitude, for the reason that her care is so often necessarily manifested in ways that are painful or irksome to their feelings. As they grow a little older, however, they often evince the most tender affection towards their nurse after they have become convalescent. At the Infirmary for Children I have frequently seen former patients show the greatest joy at visiting a nurse who has tenderly watched them in sickness.

In surgical affections a scrupulous attention to cleanliness is imperatively necessary, and it is here that the good nurse shines. In operations for stone, &c., the contrast between the progress of cases well or inefficiently tended is most remarkable. It seems almost as if the surgeon had done his part when the patient is removed from the operating table, and the rest depended on the nurse. One considerable part of the duty of an attendant for sick children consists in her keeping them well amused; and the surgeon can recognise in an instant whether in his absence they are treated with kindness, and attention paid to their little wants.

In hospitals the importance of ventilation and pure air is now understood; but in treating private cases the surgeon may have much to do in procuring a proper attention to these and other hygienic measures; many a child passes a restless night from too much clothing or deficiency of pure air. But when a lady so competent as Miss Florence Nightingale has undertaken to instruct her sex, the surgeon cannot do better than by all means in his power to enforce her counsels. On one point only I must be allowed to differ from such excellent authority, and that is in respect to opening the windows of the bedroom during the night.

DISEASES AND INJURIES OF THE HEAD.

I. CONCUSSION — FRACTURE — COMPRESSION.

Injuries to the cranium or its contents are comparatively rare in very young children, in consequence of the provision made by nature to permit the expansion of the brain; the bones being connected by yielding and somewhat elastic membranes, the force of any concussion is for the most part dissipated. No one can have witnessed the numerous falls and blows which children receive during play without being struck by their immunity from ill effects.

In the practice of a large hospital, however, numerous cases are met with of children and young lads, from three years old to the age of puberty, suffering from severe injuries to the cranium, either with or without fracture. In these cases results of the most opposite character may ensue. For instance, a child, about four years old, was knocked down by a pocket of hops falling upon its head, and died after sixty hours with stertorous breathing and convulsions. On examination, no fracture was discovered, but there were points of ecchymosis over the surface of the brain. On the other hand, injuries of very severe character occur with comparatively slight consequences. A boy, about nine years old, fell from the top of a cart, his head striking the edges of a brick. He was brought into the hospital insensible, and with stertorous breathing. The right parietal bone was fractured, and partially depressed. The bone was raised, and in about ten days he was out of danger. Again, a boy, aged twelve, while standing on a van overbalanced himself and fell, striking one side of his head. He was picked up stunned, and immediately brought to the hospital. He was insensible; the pupils natural; no symptoms of compression. There was an apparent depression near the posterior inferior angle of the parietal bone. (Was not this due to the swelling of the integuments to be hereafter spoken of?) Cold water only was applied to his head, and in six days he went out well.

Accordingly, the *Prognosis* in these cases is very uncertain; the apparent amount of the injury sustained bearing no constant relation to the issue to be anticipated.

The *Diagnosis* presents no points calling for special remark. It should be observed, however, that the apparent depression arising from the effusion of fluid under the pericranium is even more deceptive in the child than in the adult. A child between three and four years old was brought to the hospital, having a well-marked ridge, with central depression resembling that of a depressed portion of bone, just behind the parietal eminence; fluctuation, however, was evident in the centre. On inquiry, the mother stated that a week previously the child had received a blow on the head, and had not been well since. By the application of lint dipped in cold water, in the course of a week the swelling had entirely subsided, and the fluctuation had disappeared. This case had been sent to me as one of fractured skull with depression.

The *Treatment* of these cases is extremely simple. In cases of concussion, after a mercurial purge (this is the plan usually adopted, but it is possible that it might without detriment be omitted), a piece of lint dipped in cold water, or a bladder of ice applied to the head, is all that is necessary. This application seems to give great comfort, and relieves the characteristic restlessness; which is easily accounted for by the intimate communication existing between the vessels of the interior and exterior of the cranium. The rapidity with which children recover from concussion is very remarkable. This is probably due to their general vital energy, the healthy state of their viscera, and their freedom from the morbid mental excitement which so frequently retards convalescence in the adult. The ill effect of a too early education (as seen in the precocious children of the present day) shows itself in a more tardy recovery.

The surgeon should impress upon the parents' minds the probability that the child may have to recommence its education after an injury that has deprived it of consciousness. The forgetting of knowledge previously acquired, which occurs even in the adult, is much more marked in children. It is indeed, with them, the rule. I have known children, previously able to read with ease, after an accident of this kind, producing unconsciousness, lose their knowledge of the alphabet. They often seem to have to commence life over again.

In the treatment of injuries to the head attended with fracture, the same rule holds good in respect to depressed bone, as in adults; with this exception, that, owing to the absence of the diploe and of the frontal sinuses in the child, the pressure from bone driven in is more apt to be directly exerted upon the brain. When, therefore, symptoms of compression, of a doubtful character, exist, this fact should not be lost sight of, and the elevator might be immediately had recourse to in some cases in which, if the patient were an adult, it might be better to defer its use. If it be necessary to employ the trephine, it cannot be too often impressed upon the surgeon that, owing to the absence of diploe, the pin of the instrument should be very speedily withdrawn.

Children may leave their beds after injuries to the head much sooner than adults. My own experience would lead me to permit this as soon as the disposition to rise is strongly manifested. The attempt to restrain children is accompanied with more ill consequences to their tender brains than any moderate exertion, to which their natural instinct would prompt them, is likely to be. But, at the same time, prolonged quiescence would no doubt be beneficial to them, and it is of great importance to obtain it to the utmost possible extent by presenting to their notice pictures, toys, and every kind of amusement which may keep them willingly at rest. At the Infirmary for Children, it is usual to put in front of each patient a tray sliding upon the bed, on which are placed a variety of amusing objects.

A trickling of blood from the nose, accompanied with subconjunctival ecchymosis in one or both eyes, occurring immediately after the accident, is pathognomic of fracture of the orbital plate of the frontal bone, and is therefore a symptom of the most serious import. But it often happens that free epistaxis occurring a few days after the accident seems to give great relief to children suffering from symptoms of concussion. A case occurred at Guy's Hospital of a child admitted with concussion and apparent paralysis of the extremities—that is, wherever the child's legs were placed, there they would remain.

Two days after admission a sudden bleeding came on from the nose, deep sleep followed, and he woke conscious. From that time recovery took place.

The question naturally arises: Can we artificially obtain the same benefit by the abstraction of blood from the arm or temples? I should unhesitatingly answer, No. During my first years of practice, bleeding from the jugular vein, or temporal artery, was almost universal in the earlier stages of concussion. The results were unequivocally disastrous. And when it is remembered how ill children, under all circumstances, bear the abstraction of blood, this is not to be wondered at. It should never be forgotten that these tender plants require all their sap for the nutrition of their tissues, in towns especially.

In collapse, it is better to abstain from the use of stimulants, and let recovery take place spontaneously. There is always much danger of an excessive reaction in the young.

CHAPTER II.

DISEASES AND INJURIES OF THE FACE.

I. WOUNDS OF THE FACE.

IT is only necessary to allude to these on account of the especial care requisite for their treatment in children. Since the cicatrix grows with their growth, the scar should be kept as small as possible by the most accurate adaptation of the edges of the wound. In incised wounds there is of course no doubt in respect to the treatment, but I have also obtained the happiest results in various forms of lacerated wounds by bringing the edges into close union. This is best done in every case by using one of the needles depicted in 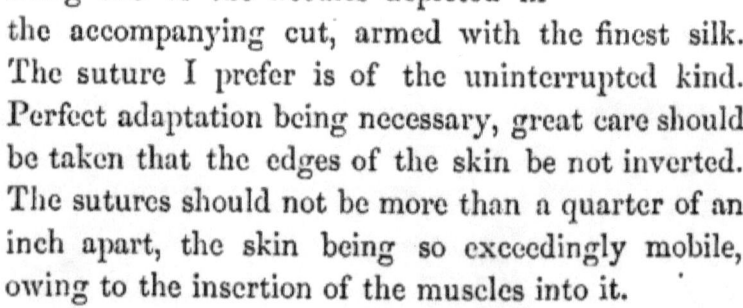 the accompanying cut, armed with the finest silk. The suture I prefer is of the uninterrupted kind. Perfect adaptation being necessary, great care should be taken that the edges of the skin be not inverted. The sutures should not be more than a quarter of an inch apart, the skin being so exceedingly mobile, owing to the insertion of the muscles into it.

Even should suppuration occur, it is better not to remove the sutures, unless there arise constitutional disturbance or a great amount of swelling. If the sutures tear away, and suppuration be established, the edges of the wound being separated, it will be advisable to reapply the sutures, and bring the edges together. This may be done more than once, and it is the only means by which disfigurement can be avoided. In the eyelids or lips this is of greater importance than in other parts of the face. I have met with a case in which the lobule of the ear was almost entirely separated, but by careful adaptation perfect re-union was obtained.

Metallic sutures I have frequently used, but in cases of this kind I give the preference to silk. It is more easily applied, and does not create any greater irritation. Indeed I have seen as much irritation from the metal as from silk in various parts of the body.

It is of course advisable to allow the bleeding to cease before applying the sutures, and to avoid ligatures, if possible, as these impede the union.

The same principles apply to all operations on the face, in children, such as the removal of cysts, &c. A careful adaptation of the edges of the wound is necessary, with a view of obtaining speedy union, and avoiding a scar.

Whilst speaking of injuries to the face, I may mention that a condition alarming to the parents is sometimes produced by a violent fit of hooping cough rupturing a sub-conjunctival vessel, and causing an effusion of blood over the surface of the eye. This becomes absorbed in a short time.

II. DISEASES AND INJURIES OF THE NOSE.

Fracture of the nose may occur in children from falls or from the blow of a cricket ball, &c. Any displacement should be immediately rectified, however slight it may appear, as what may be a very trifling deformity in the child may become a serious defect in the man. This arises from the mode of growth of the parts, as shown by Mr. Hilton, the expansion of the sphenoid bone pushing forward the vomer, and the septum nasi, and subsequently also the nasal bones. Any injury, therefore, to the septum or nasal bones, attended with displacement to either side, would necessarily result in a progressive deformity, since the bone would continue to grow in the abnormal direction.

If driven in, these bones are easily replaced by passing a pair of dressing forceps gently up the nostril of the injured side. By carefully manipulating with the finger on the outside, and the forceps within, the displaced bone can generally be returned to its position. Once being rightly placed, the parts will remain in situ, no splints or other apparatus being necessary.

III. FOREIGN BODIES IN THE NOSE.

No one who has seen children playing with pebbles, marbles, small pieces of wood, &c., and who remembers their disposition to place these various articles in any of the inlets of the body, could fail to be surprised if the nose should enjoy an immunity from their presence, notwithstanding the great sensitiveness of its mucous lining, and the narrowness of its aperture. Such accidents generally happen without the cognizance of the parents or nurse; the child does not complain, and it is only when the foreign body begins to excite certain marked symptoms that the surgeon is consulted. The child is generally observed to have the nose a little swollen, tender to the touch, and more or less red. These symptoms, in the course of a few days, pass off, leaving only a discharge, more or less offensive, or even free from odour, which may exist for weeks, months, or even years, without its true cause being detected. A child was brought to me who had suffered from so-called ozæna for eight months. The presence of some foreign body had been suspected by the medical attendant, and when the patient was brought to me I carefully examined each nostril with the lamp and speculum, but we were both unable to discover any. My opinion was, that some foreign body was present, on account of the discharge being confined to one nostril. We recommended the constant injection of warm water, and at the end of three months a small piece of cane came away. The lamp I use for all purposes in which a strong light is required, is depicted in the accompanying woodcut. It is a gas lamp, with a reflector, and may be used either with or without a lens.

The *Diagnosis* is by no means easy, unless the foreign body can be seen, which is generally extremely difficult, owing to the rapid swelling of the very vascular mucous membrane. The discharge being confined to one nostril, however, is a very important indication; but it is not constant, inasmuch as the

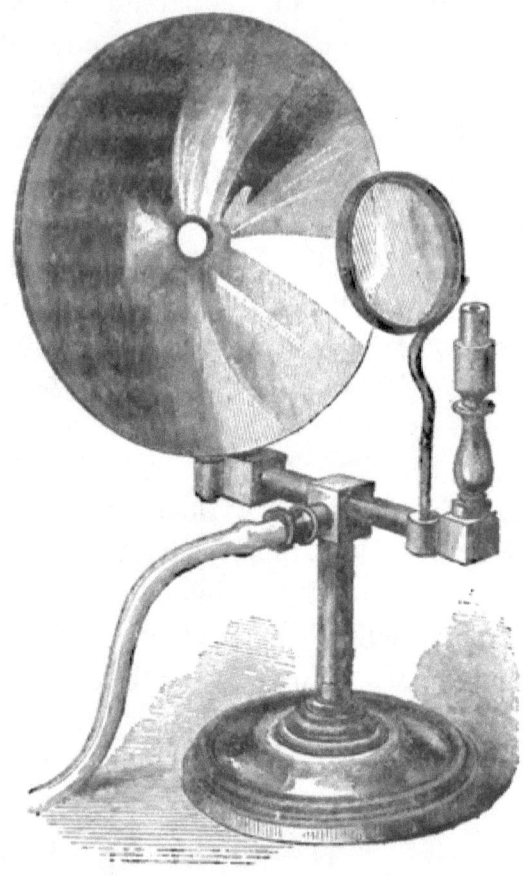

septum may be ulcerated through. Of course the surgeon will distinguish between these foreign substances and the firm masses of inspissated mucus which may accompany ozæna.

In the *Treatment* of these cases syringing is often very useful, but it should be had recourse to only if the foreign body cannot be seized and brought away by the forceps, or by means of a scoop insinuated behind it. In this respect, the treatment is the opposite to that which is suitable in the case of foreign bodies in the meatus of the ear, in which the syringe alone should be used. The difference in the anatomical relations of the parts accounts for this difference of treatment. Should the mucous membrane be greatly swollen, we must wait. The swelling subsides in a few days.

IV. EPISTAXIS.

Bleeding from the nose in children is perhaps one of the commonest of accidents. On the slightest injury, or even without any injury at all, a loss of some ounces will occur, and this loss will ensue repeatedly, without producing much effect upon the general health. Two questions suggest themselves here: first, Why does the bleeding so frequently recur? and, secondly, Why do the children suffer so little?

Perhaps no positive answers can be given to these questions; but some probable causes may be assigned. Thus, in consequence of the very active circulation in the brain, enclosed as it is in an unyielding case, it may be necessary in some children that a means should be provided for affording an escape for a temporary superabundance of blood. Under excitement, the amount of blood in the brain is greatly increased, and it is possible that disastrous effects might result from this, if it were not for the free exit afforded by the various foramina of the skull, of which the most important is the foramen through which, *in children*, the blood passes into the nose. In after life, when the passions become more amenable to the reasoning powers, this foramen becomes closed. The peculiar construction of the Schneiderian membrane affords a means for the ready transudation of, at all events, the more fluid portions of the blood. And this may be in part the reason why children suffer so little from the effects of this hæmorrhage—viz., that it is not strictly blood, but an altered and diluted fluid that escapes; coupled with the fact that the loss is from the veins, and not from the arteries.

In some cases the tendency to epistaxis is hereditary; or it may, I think, be due to accidental occurrences depressing the mother's strength during pregnancy. I have met with a case in which one child in a family of three had a strongly marked tendency to loss of blood, especially by the nose. There was no such tendency in the parents or their families, but the mother during the sixth month of pregnancy had met with a fall, occasioning her a considerable loss of blood.

Treatment.—In ordinary cases no attempts should be made to stop the hæmorrhage: they are for the most part quite ineffectual. The patient should be freely exposed to the air, and should, above all, be kept standing. In this posture fainting is induced much more readily than while recumbent, and consequently the hæmorrhage more quickly ceases. Cold sponging may be used to the nose, and no opposition need be offered to any harmless domestic remedies. Ordinary styptics are of no avail. If life should seem to be endangered, of which I have never known a case (and it is remarkable that fatal hæmorrhages in children take place from various other parts without implicating the nostrils), any means that promise a chance of success may be had recourse to: the most efficient would probably be plugging the nostrils, or the application of a weak solution of perchloride of iron; but I have never needed to employ either.

V. CHRONIC THICKENING OF THE SCHNEIDERIAN MEMBRANE.

We are often consulted by the parents of delicate little children on account of a "stuffiness" of the nose, accompanied with superficial ulceration of the orifice of the nostrils. Upon looking into the nasal passages, we observe a thickening of the Schneiderian membrane covering the septum, and almost invariably confined to that part. This is generally accompanied with slight redness and a semi-purulent secretion. It is the disease called "strumous disease of the Schneiderian membrane" by Sir Benjamin Brodie, who first pointed out the distinction between it and polypus, for which it is very often mistaken. It is of great importance to distinguish between these affections, as any attempt at removal is attended with the unpleasant result of entire failure.

Chronic thickening is distinguished from polypus by the fact of its occurring on the septum, whereas polypi invariably grow from some one of the turbinated bones. The surface of the polypus is smooth and shining; that of the diseased membrane rough and irregular. The discharge is thicker than that from polypus; moreover, the swelling does not alter its position. If the finger or a probe is passed up the nostril, the thickened membrane is found to be sensitive to touch, which polypus is not; and it has a broad base, whereas polypus is pedunculated.

The *Treatment* of these cases is to improve the health. If the circumstances of the patient admit, change of air, either to the sea-side in the autumn, or into the country in the spring and summer, is by far the best remedy; the nitrate of silver, either in stick or in strong solution (3j ad ʒj), being at the same time applied.

Amongst the poorer classes, in whom this disease is much the more frequent, the attempt should be made to obtain a change of air. A milk diet, with meat two or three times a week, and no more, should be enjoined; and for medicine, oleum morrhuæ and vinum ferri, of each ʒj twice a day, may be given after a meal. The local application of the nitrate of silver, as above mentioned, may assist the cure.

VI. POLYPUS OF THE NOSE.

There is scarcely any cavity of the body more prone, in children, to polypoid growths than the nose. These give rise to much uneasiness, though without actual pain. They are of various kinds, but it is not my intention to describe them separately; for all practical purposes, the hard and the soft polypi may be treated under one head.

A child is brought with a pedunculated shining body appearing in one nostril. It has been supposed to have had a cold for some time, with more or less difficulty of breathing. The "stuffiness" is not so marked as in the case of chronic thickening, and at times the patient will breathe almost naturally; its uneasiness being much relieved, owing to the polypus ascending into the upper part of the meatus, and leaving the air-passage free. On violently blowing the nose, however, it may again be made to appear near the orifice, and the obstruction returns, contrary to the usual effect of blowing the nose. There is a constant secretion of tenacious mucus. Polypus can only be mistaken for the disease last mentioned; but, unlike that, it occurs for the most part in healthy children.

The *Treatment* is, of course, removal. I have quite abandoned the use of the forceps, and find the ordinary wire snare, shown in the accompanying woodcut, by far the most useful instrument. Previous to its application, I take the precaution of ascertaining, if possible, by means of a probe, the seat of the attachment of the polypus. The wire should then be carefully placed around the pedicle as near the root as it can be applied. Slight bleeding follows the removal, and the child experiences immense relief after a short time. The relief is not immediate, owing to a coagulum of blood blocking up the nostril. The friends must not expect a complete immunity from the return of the disease; a second or third operation is frequently necessary, however complete the first removal may appear. The difficulty of this operation consists in dexterously encircling the polypus within the noose.

VII. STRUMOUS ULCERATION OF THE FACE.

Children are sometimes brought to us with a superficial ulceration of some part of the body, but more frequently of one or other cheek; irregular in shape; apparently inclined to heal in one part while spreading in another; the edges thin and eroding; no appearance of granulations; the surface of the sore smooth and shining, as well as the surrounding integument; the discharge watery, rarely purulent; a part of the sore has frequently scabbed over; and if a part has healed, there is a puckering of the new skin. These children are generally delicate, fine-skinned, with long eyelashes, and otherwise very pretty; and if carefully examined, some few enlarged glands may be discovered in the neck. Now and then this affection is accompanied with disease of one or more of the metatarsal or metacarpal bones, or with enlargement of the extremities of the radius and ulna, or the child is knock-kneed; though it may occur without any other symptom of scrofula. The mother generally says that all sorts of remedies have been applied to the ulceration.

By some this disease would be described as lupus, and I am not prepared to say that it is not analogous to that affection in the adult. This I know, that any attempt to cure it by escharotics, as in that more troublesome disease, is generally attended with the worst results.

The *Prognosis* in these cases is very unfavourable as far as regards a speedy cure; but there is no danger to life. It is a disease not commonly recognised, and is frequently taken for an affection of the skin—at least the plans of treatment adopted would seem to indicate this.

In the *Treatment*, the use of irritating applications to the sore should be avoided. A small piece of lint, dipped in water, cut to the size of the wound, with a piece of thin gutta percha over it, is all that is advisable in the way of external application. Attention should be paid to the general health. The universal remedies of pure air and a wholesome diet, consisting largely of milk, should be enforced. If no improvement occurs in the appearance of the ulceration, the following formula may be found useful:—

℞ Tinct. cinchonæ . . . . ♏ xv.
— rhei . . . . ♏ vij.
Liq. hydr. bichlorid. . . ♏ x.
Aq. destillat. . . . . . ad ℨj.

The above is the dose for a child five years old.

I have also found iron, in the shape of the sulphate, the sesquichloride, or the iodide, useful. And I may here observe that the old-fashioned vinum ferri, which is so largely used in children's diseases, may owe its good effects more to the amount of stimulant than to the iron it contains. A very elegant mode of exhibiting this medicine to children is to be found in the "steel biscuits," which I have known them take when the drug has been refused in all other forms. The usual anti-strumous remedies of cod-liver oil, quinine, &c., may be given to vary the treatment.

Case.—H. M., aged nine years, was admitted into Guy's Hospital on 7th December, 1842. Ever since he was a year and a-half old, he had been subject to attacks of superficial ulceration on various parts of his body. He has had ulceration of the face for about three years, previous to which time there had been a similar affection, first of the left and then of the right shoulder. On admission, there was a patch of ulceration, of the character I have described, on the right side of the face, extending over the whole surface of the masseter muscle; on the right shoulder there was also a patch of ulceration, irregular in form, extending quite across the deltoid. The treatment consisted in applying a rag, dipped in yellow wash, to the parts affected, and the exhibition of one-twelfth of a grain of bichloride of mercury in sarsaparilla three times a-day. On the 3rd of January, the lotion was discontinued, and water-dressing applied, and subsequently the iodide of iron was given; the lad went out well on the 16th of May.

VIII. CANCRUM ORIS.

Of all the awful diseases which the surgeon is called upon to treat, occurring in children of tender age,

Cancrum oris is the most frightful. In its more advanced stages, the patient presents one of the most hideous spectacles that can be imagined, and thankful indeed must many a parent have been when death has given the sufferer release. By some authors this affection is wrongly described as "gangrenous stomatitis," which, though a severe, is nevertheless a different disease. Cancrum oris very frequently follows a depressing malady, or one of the eruptive fevers. It may, however, occur in children confined to badly ventilated dwellings, where poverty, with its attendants of dirt and semi-starvation, is prevalent. I am, however, disposed to believe that though these may be set down as predisposing causes, the indiscriminate use of mercurial powders, a portion of which very frequently is retained in the mouth of young children, is, in the larger proportion of cases, the exciting cause. I have never yet taken the history of one of these cases, without finding that mercurial powders, generally procured from the chemist, have been given. At the same time, mercury is so frequently given to children in all kinds of diseases, that no positive conclusion can be drawn.

This disease commences by a redness, swelling, and a peculiar hardness at or above the angle of the mouth, and upon looking within (if we can open the mouth sufficiently), we observe an ulceration on the surface of the mucous membrane, of a greater or less extent, sometimes as large as a shilling. This latter may be considered as the first morbid alteration. The surrounding mucous membrane is swollen, shining, and pouring out an abundant secretion. As the swelling increases, the saliva dribbles out of the mouth, and sometimes in less than twenty-four hours the reddened skin without has given way, and formed an excavated gangrenous ulcer; for the whole tissues of the cheek are involved in this frightful malady. The ulcer presents an appearance as if a piece had been punched out.

As the disease progresses, which it does most rapidly, the whole cheek and side of the face become swollen, and assume a peculiar waxy colour; the part immediately surrounding the ulcer is extremely hard, and quickly becomes involved in the gangrenous destruction; making another considerable addition to the already gaping wound. The lips are generally much swollen, and the opposite side of the face becomes infiltrated with serum. Another day passes, and again the hardened integument around the sore warns the surgeon of the fearful destruction which is about to ensue. At length a yawning gulf is seen, at the bottom of which may frequently be observed the blackened teeth, and the upper and lower jaw denuded of the periosteum, the pharynx being frequently exposed. At the commencement of the disease the pain is extreme; subsequently drowsiness supervenes, and affords some relief. The constitutional disturbance is very variable; some children seem to suffer much less than others. The discharge is composed of shreds of sloughing tissue, pus, and blood. The fœtor is intense. At Guy's Hospital, a year never passes without one or two of these terrible cases being admitted.

Case 1.—H. T., aged twenty-one months, was admitted into Guy's Hospital, under Mr. Hilton's care, in October, 1855. The mother stated that the child was in good health until fourteen days before, when it had a mild attack of varicella, for which she procured some powders at a chemist's. Nine days before admission, swelling of the face came on, and at that time there was pain, with ulceration of the mouth and gums. On admission, the whole of the right cheek was destroyed, gangrenous, and emitting a fœtid odour. The gangrene continued to spread, until the pharynx was exposed, as well as the bones of the upper maxilla. The nose was also destroyed. The child lived altogether eighteen days, an unusually long time. In this case the vessels going to and coming from the diseased part were carefully examined, but nothing was discernible. The facial artery, where it entered the sloughing mass, was closed by a natural clot.

Case 2.—M. A. W., aged three and a-half years, was admitted into Guy's Hospital, under Dr. Gull's care, May 17th, 1857. She had always been sickly, and ever since birth had had a frequent cough. A year ago she had hooping cough, which lasted three months, since which time she had been better, and was considered to be in tolerable health up to three weeks ago, when she was again ill with her cough. Calomel powders were given. A week before admission the mother noticed a black spot on the inside of the lip, on the right cheek, and on the gum, and soon afterwards the teeth fell out. The child had had no

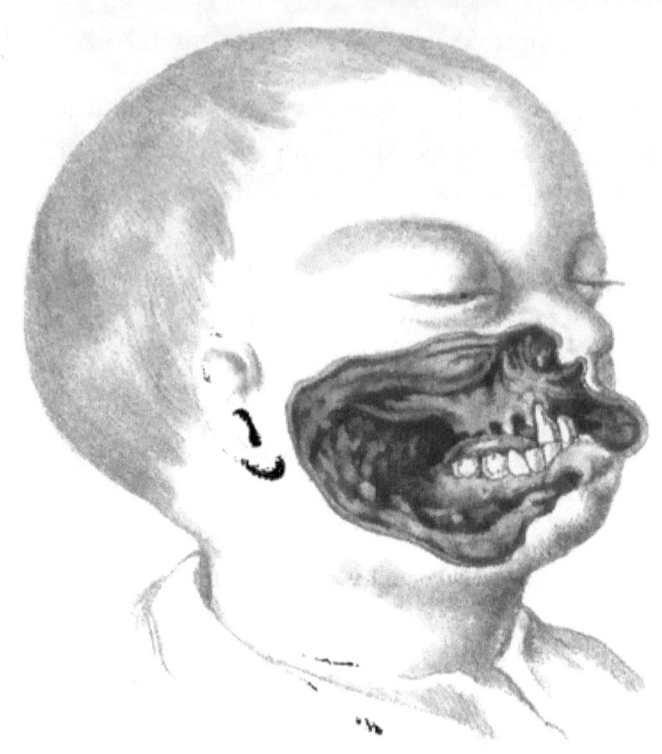

fever, measles, or scarlatina. On admission there was a large black slough on the right cheek, going through into the mouth and involving the gums. At the end of a week the child died. On the post mortem there was found inflammation of the bronchi; all of them, even to the smallest branches, being filled with a creamy mucus. The appearance of the lungs was such as is found in the broncho-pneumonia of children.

In respect to the *Diagnosis*, there is no disease with which it can be confounded; except, in its early stages, with gangrenous stomatitis, which is a much milder affection. The following case is an example of it:—C. H., aged thirteen months, admitted into Guy's Hospital in January, 1856, under Dr. Addison's care. Six weeks before admission had measles, followed by hooping cough; blisters appeared on the tongue and mouth; the gums and the lips soon became affected with gangrenous ulceration, but no perforation took place through the cheek. The child quickly sank.

My own view is, that these are generally distinct affections; but in the opinion of many the difference is only one of degree.

In respect of the *Treatment*, the milder disease last mentioned, though frequently fatal, is yet often amenable to stimulants, chlorate of potash, and milk diet. True cancrum oris, on the other hand, is nearly always fatal. The best treatment known at present seems to be the application of strong nitric acid (under chloroform, of course), until the gangrene is arrested. Lint dipped in chloride of soda lotion (ʒj. of the liquor sodæ chlorinat. to ℥j. of water) should be kept constantly applied. The internal treatment consists in steadily supporting the powers with milk and beef-tea, with an unlimited quantity of wine. The supposed specific remedy—chlorate of potash—appears, from my experience, to be utterly useless.

If recovery takes place, so far as I have seen, the contraction of the cicatrix produces the most frightful deformity of the face; the mouth is sometimes reduced to such a size as scarcely to admit a quill. Should we be fortunate enough to arrest the progress of the disease, part of the chasm formed by the gangrene may be closed by a plastic operation.

Children who have suffered from the milder affection—stomatitis gangrenosa—are sometimes brought to the surgeon unable fully to open their mouths, from contraction of the buccal mucous membrane, and adhesion of it to the gums of the upper and lower jaw. I have a patient, aged five years, at present under my care, in whom this condition exists on both sides; the child is only able to open its mouth to a very slight extent, and on introducing the finger a firm band can be felt extending from the upper to the lower jaw. The affection may be recognised at once by the peculiar puckering which takes place in the cheeks when the child attempts to open its mouth. As the child grows, it appears to me probable that the powerful action of the depressors of the jaw may be sufficient to stretch the new tissue, and render unnecessary any operative proceedings. But as yet I am not acquainted with any observations bearing upon the point.

CHAPTER III.

DISEASES OF THE MOUTH AND TONSILS.

I. HARE LIP.

NO one can have witnessed without sympathy the anxiety evinced by many mothers that the new-born child should be perfect, or have failed to share their distress when the infant is presented to their notice with the deformity called Hare Lip. And it becomes the bounden duty of the surgeon to bring into exercise all the resources of his art to remedy the painful defect. There are various forms of this affection. Hare lip may be either single or double—that is, affecting one side or both; it may be confined to the lip, or implicate the hard and soft palate to any extent. In some cases it extends into the meatus of the nose, widening the alæ, and so giving a very disagreeable appearance to the face. Moreover, one part of the jaw and lip seem to be on a plane behind the other; what might be fairly described as the inter-maxillary bone, immediately below the septum of the nose, appearing to project inordinately, especially if the hare lip is double: and this gives rise to a great deal of inconvenience in the operation. The accompanying woodcuts represent the usual forms of single and double hare lip.

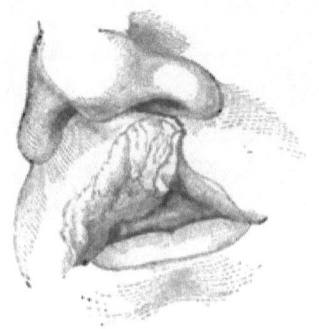 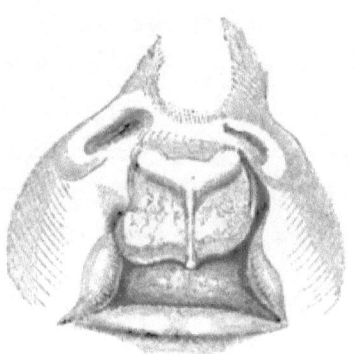

Like some other deformities, hare lip often runs in families; and I have met with instances in which several children of the same parents have been so affected. In all these cases it has happened that the father has been the subject of hare lip, and not the mother. In one family of nine children the first two had hare lip, and imperforate rectum; the third had imperforate rectum and defective palate, but no hare lip; the fourth, fifth, sixth, seventh, and eighth, were free from external deformity; the ninth had imperforate rectum. At this time the operation for hare lip was performed on the father, and there have been two healthy children since.

The *Treatment* of these cases is one of the most interesting operations connected with the surgery of children. An important point in connexion with it is the time at which it should be performed. Should the infant immediately after birth be subjected to the tender mercies of the surgeon, or should time be allowed for the child to acquire more constitutional vigour? My experience is decidedly in favour of operating at the earliest possible period; remembering, however, that loss of blood, in very early infancy, may be attended with the worst consequences. I should have no hesitation in operating immediately after birth, so as to avoid the shock to the mother occasioned by the sight of the child. The operation has been performed successfully within seven hours after birth. But the possibility of a fatal issue must not be forgotten.

The principal reason for which I prefer to incur the greater risk from hæmorrhage in very early life is, that the child may be enabled to suck. The danger from the operation is less than that from the imperfect nutrition which results from the deformity. By operating early there can be little doubt a certain per centage of deaths from this cause might be avoided. In the following cases, it is probable that death ensued mainly from want of proper nourishment.

Case 1.—J. W., aged six weeks, was admitted into Guy's Hospital, July, 1854, to be operated on for hare lip. Like many of the class, it had been unable to suck efficiently, and was becoming emaciated. The usual operation was performed on August 9th. There was no excessive loss of blood, but the child died the next day with convulsions and diarrhœa.

Case 2.—W. H., aged eleven days, was admitted under my care into Guy's Hospital, August 6th, 1856, with hare lip and fissure of the hard and soft palate. The fissure was very wide, the child badly nourished, and unable to take the breast, and I considered it a most unfavourable case for operation. The usual proceeding was adopted, and for two days the child took the breast well. It then, however, began to reject the milk, and died in four days afterwards.

By operating very early, not only is the risk of starvation avoided, but the deformity is much less.

If the parents object to the operation being performed before the period of dentition, it is then generally said to be advisable to postpone it to the second year; but I have performed it at all ages with success. Of course the time when a tooth was actually penetrating the gum would be avoided.

Operation.—In single hare lip, the child being, of course, placed under the influence of chloroform, the head should be steadied by an assistant, and one or two others should be ready to seize the divided lip, so as to compress the vessels. Dexterity, on the part both of the operator and those with him, is very important to restrain the hæmorrhage from the coronary arteries, which, every assistant should remember, run, not beneath the skin, but immediately beneath the mucous membrane of the lip. It signifies but little whether the patient is placed with his head or his feet towards the operator.

The first thing to do is to free the lip from the gum to the extent of half-an-inch on each side; one edge of the fissure should then be seized with a pair of artery forceps, and a sharp-pointed knife should be inserted just above the upper angle of the aperture, pushed through the substance and mucous membrane of the lip, and carried downwards to its free edge, where a slightly curved direction towards the fissure should be given. Care should be taken to remove a sufficient quantity, so that the raw edge may extend the full depth of the natural thickness of the lip. One assistant should now quickly seize the divided edge, and compress the coronary artery. A corresponding incision should then be made on the opposite side, and another assistant should immediately seize the lip and compress the vessel. The great art now consists in accurately adapting to each other the edges of the wound thus made by the surgeon, and here the greatest nicety is required. As a general rule I discard the pins *in toto*. For this reason: In all cases in which I have seen them used, there has been almost invariably a scar left at their points of entrance and exit. The only exception that I could allow would be, where there was very great tension upon the lip, in which case one pin might be kept in for twenty-four hours. The plan that I always adopt is the same as that which I have recommended for incised wounds of the face—viz., the uninterrupted suture. The needle should first be inserted at the lower part, so as to secure the most absolutely exact adjustment of the prolabium, and the sutures carried upwards. In introducing the stitches, great care should be taken that they include a considerable portion of the substance of the lip. I make it my object to transfix the coronary artery without piercing the mucous membrane. If the latter be done, the child is made fidgety by the presence of the suture in the mouth; and if a small portion only of the lip be taken up, the surface becomes united without the deeper parts undergoing union. Acupressure, therefore, is by no means so novel a plan for restraining hæmorrhage as has been supposed. I have used it in this way for many years. Ligatures, of course, are out of the question. The only other precaution that needs to be taken, is to guard against the turning in of the edges of the wound, which results in partial non-union. The threads may be removed any time after the fifth or sixth day. Mr. Butcher, of Dublin, recommends the use of a pair of sharp scissors, instead of a knife, for paring the edges.

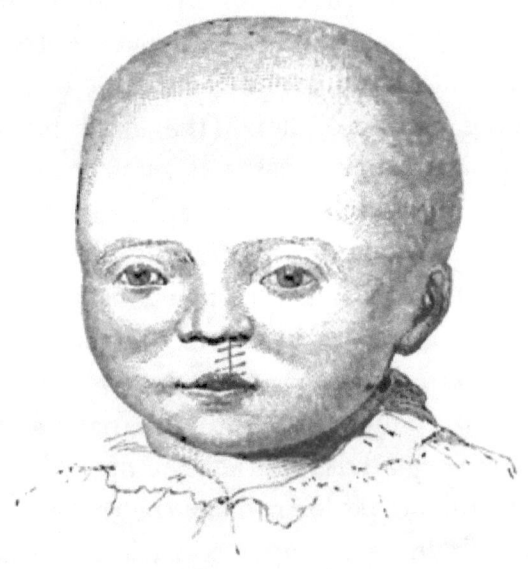

The accompanying woodcut is taken from a little patient of Mr. Cock's, that was operated on in June of the present year, and represents the state of the parts immediately after the operation.

In double hare lip, the same operation must be performed on each side, and at the same time. The shortness of the portion of lip covering the intermaxillary bone renders it necessary that the edges of the two lateral portions shall be brought into union inferiorly; the central portion (its edges, of course, being pared) serves to fill up the gap between the two. In some cases it is advisable to remove the projecting portion of bone. It should always be broken away, if possible, with the fingers, not divided with the cutting pliers, unless it be absolutely necessary. I once saw an oozing of blood continue for two days from the surface of a bone so divided; the child died, apparently in consequence. If the cutting pliers are used, the hot iron had better be applied to the bleeding surface of bone (the child being under chloroform). In some cases it may be sufficient to press the projecting bone backwards with the fingers, without removing it; in others, where the projection is very great, the skin which has covered it may form an useful septum to the nose, the bone being removed entirely.

In hare lip complicated with fissured palate, hard or soft, it is very advisable that the operation for the former should be performed at the earliest possible time, as it induces an approximation of the parts behind. In some of these cases the fissure of the palate will almost entirely disappear.

If union should not occur at first, I still think it advisable to keep the sutures in their position, or to apply fresh ones of the interrupted kind in their place, if the tension can be thus relieved. It is astonishing how often an unpromising case will, by this means, gratify us with a quite unexpected success.

It will be seen that I dispense with the ingenious apparatus (consisting of a spring round the head), suggested by one of Mr. Ferguson's patients. I have never seen a case in which I thought it advisable to use it.

II. TONGUE-TIE.

The surgeon is occasionally consulted by an anxious nurse, who supposes that the infant under her charge is what is called tongue-tied, because it makes a peculiar noise in the act of taking the breast, or is apparently unable to seize the nipple with full effect. I have observed the greatest anxiety evinced by these old ladies when the child is a girl; I presume, lest the unruly member should not have full play in after life. In eight cases out of ten, there is not the least necessity for doing anything. Now and then one sees an excessive folding of the mucous membrane forming the frænum, which generally, however, stretches in after life without surgical interference. But should the surgeon deem it advisable to do anything, he may make the slightest snip with a pair of probe-pointed scissors, in a direction backwards (not upwards towards the tongue), and then press the tongue back; by this means tearing the frænum rather than cutting it. To put the frænum on the stretch, it is sufficient to make the child cry.

III. RANULA (SO CALLED.)

The disease called ranula, that is to say, dilatation of the submaxillary or sublingual ducts from obstruction at their orifice, so far as my experience extends, is not met with in children. Indeed, I doubt very much whether it exists at all. Judging from analogy it appears improbable, for why should not the same thing occur in Stenson's duct? I have also, at the very time that ranula has been supposed to exist, passed a hair probe along the duct of the sublingual gland; proving thereby that this swelling has not any connexion with closure of the duct. What then, my readers may ask, are the swellings that form under one side of the tongue, apparently like thin cysts, transparent, and filled with a glairy fluid, with one or two tortuous veins meandering over them? They are nothing more than obstructed mucous follicles, corresponding to the cysts which sometimes may be seen in other parts of the mouth from obstruction of the follicles; and they are quite unconnected with the duct. The accompanying woodcut represents the usual appearance of ranula.

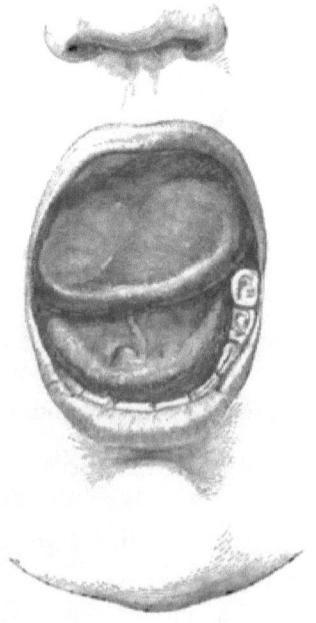

When first seen, these

swellings vary in size from that of a small marble, to that of a bantam's egg. The child, therefore, has evidently suffered from them for some time before it complains, and the mother's attention is usually first drawn to them by some alteration either in speaking or in swallowing. The saliva also sometimes flows in excessive quantities from the mouth. The cyst, when confined to one side, may so far enlarge as to extend across the mesian line, and produce an appearance as if there were two. It may even be difficult to decide from which side the cyst has grown.

Simple as this disease may be in its pathology, its curative *Treatment* is by no means easy. When possible—that is, when the cysts are in any part of the mouth, except beneath the tongue—there is no doubt that the most effectual cure is to make an incision through the mucous membrane, and remove the cyst entire. Under the tongue, this plan of treatment is practically out of the question. We then have to adopt some other means for getting rid of the secreting membrane of the follicle. For, so long as that remains unremoved, unaltered in character, and capable of secreting mucus, so long is there a liability to a return of the disease. It is not, therefore, sufficient simply to lay open these cysts and allow the discharge of their white-of-egg-like fluid. Laying hold of the cyst with a pair of forceps, and snipping out a piece, is sometimes found to be sufficient; but it cannot be relied upon, as the edges of the cyst will re-unite, and a fresh secretion occur. This will often be the case, even when the nitrate of silver is applied to the interior of the cyst after excision of a portion of its walls. Various other plans of treatment I have adopted; the one which I now employ, and which I believe is as successful as any, is that of passing a curved needle, armed with a double thread, through the cyst, and letting the thread remain in as a seton for four or five days till suppuration is fairly established. I have seldom failed to obtain a radical cure by this means, though I cannot say my success has been invariable.

Care should be taken that the cyst is fairly transfixed by the needle. I have seen the cyst tapped, and then, owing to the flinching of the child, the needle passed between the cyst and the mucous membrane. This is especially apt to occur when the cyst is small: when it is large, all the parts are tense, and such an accident is hardly possible.

IV. CLEFT PALATE.

Though an affection of early life (being a congenital malformation), cleft palate can hardly be included among the surgical diseases of childhood. It is useless to attempt any operation for its relief until the patient has acquired moral self-control; and its treatment, therefore, belongs to the period of adult life.

The only alleviation that can be advised during childhood, is the use of an artificial diaphragm, the benefit of which, however, is very questionable.

V. DISEASES OF THE GUM.

On looking into the mouth of children, we frequently find on the gum little fistulous openings connected with decayed teeth. Of course, the cure for these would be the removal of the teeth, but it is better to leave them unhealed. They are attended with no ill results, and it is undesirable to remove any of the first set of teeth. The only circumstances in which it is advisable to do so, are when the fangs are exposed, and by pressing on the mucous membrane of the cheek, cause its ulceration.

Occasionally we find on the gums a small vascular tumour, called an epulis. I have not met with this under nine or ten years of age. It is to be treated in the same way as in adult life; viz., by entire removal of the diseased soft parts, with a thin plate of the alveolar process.

VI. TUMOUR OF THE ANTRUM.

Tumours are sometimes developed in the antrum, or between the alveoli of one or other superior maxillary bone of children; but they are by no means common. The following is a good illustration of their course and treatment. M. J., aged seven years, was admitted into Guy's Hospital, under the care of the late Mr. Aston Key, on March 1st, 1847. Eleven months previously her face began to swell, and became painful, the swelling continuing; three months afterwards, she came into the hospital; the swelling was punctured from within the mouth, and found to be a small cyst, containing a limpid fluid, which, upon its contents being evacuated, granulated and filled up. Since that time, however, it had been increasing in size, but had never been painful. Upon admission she was a healthy-looking, florid child. There was a swelling about the size of a small orange on the left cheek. It was rather hard, and very seldom gave pain except at night. The mouth was drawn towards the left side by the tension of the skin. Mr. Key stated the tumour to be, in his opinion, of fibro-scirrhous character, probably situated between the alveoli of the upper jaw, and extending into the antrum. Fourteen days afterwards the mass was removed. An incision was made from the angle of the mouth, upwards, to the edge of the malar bone, and the upper flap of integument dissected up. The tumour thus exposed was in part removed by the knife and forceps, and the rest destroyed by nitric acid. It was found to be excessively vascular, and had somewhat of a fungoid character. In sixteen days, the wound of the face had quite healed, though there was still slight swelling. A few days afterwards, she left the hospital apparently well.

VII. BONY TUMOUR OF THE UPPER OR LOWER JAW.

We are now and then consulted for a swelling in the jaws of children, which seems to be an expansion of the outer table of the bone. In these tumours, the second teeth are sometimes found, undeveloped. The child is said to have had a swelling gradually forming, unattended with much pain: it is smooth and globular; very firm, and evidently of a bony character. The teeth are irregular and deficient. The treatment is very satisfactory. If the outer shell of bone be removed, and the contained teeth taken away, the surface will granulate and heal.

VIII. STRUMOUS DISEASE OF THE HARD PALATE.

There is a peculiar affection of the hard palate, not noticed in works on surgery, of which I have seen two instances, one of which will illustrate the nature of the disease.

C. W., living at Deptford, a strumous-looking girl, aged eight years, the daughter of healthy parents, was brought to me in May, 1858. The mother says she knows very little about the case, as the child had not complained of any pain; but three months before, she told her mother that there was a hole in the roof of her mouth. There had not been any previous uneasiness, nor had the child been aware that there was anything wrong. When I first saw her, as in the only other case I have seen, there was a ragged irregular ulceration of the hard palate behind the incisor teeth of the upper jaw, exposing the bone. On examination, not only was the bone found to be bare, but there was an opening into the nose. There was an offensive discharge, but no pain. The treatment adopted was the administration of cod-liver oil twice a day, and chlorate of potash in infusion of bark. In each of the two cases pieces of bone came away; the openings closed, and the whole healed up entirely. No local application was employed.

IX. DISEASE OF THE TONSILS.

Acute inflammation of the tonsils, commonly called cynanche tonsillaris, is a disease rarely met with in children. It seldom occurs under the age of puberty; but the slow and gradual enlargement of these organs is very frequent. This generally occurs in delicate children living in low, damp situations, and is, therefore, very frequent on the south side of this metropolis. The disease itself is simple hypertrophy of the gland substance, exhibiting beneath the microscope nothing but a mass of fibrous tissue. When the child is brought to us we may generally at once recognise the nature of the malady. The little patient walks up to us with an almost silly expression of countenance, due to the mouth being slightly open. It is not that the children are otherwise than intelligent, but the obstruction to the passage of air through the nostrils induces them to adopt this means of increasing the supply. The mothers complain that every little cold they get " flies to the throat." There is more or less difficulty in breathing, and sometimes the passage of the air into the lungs is so obstructed as to cause a falling in of the walls of the chest. The woodcut (p. 44) shows the open mouth and the pigeon-breast; it was taken from a patient of mine a few weeks ago. Another pretty constant symptom is snoring at night, and very frequently the mother complains that the child is deaf. Now I have no hesitation in saying that this deafness does not arise from the enlargement of the tonsils, otherwise it should occur in every case of such enlargement. Any one who has seen much of this disease in children must have observed very many cases unaccompanied by deafness. This has been attributed to the enlargement taking different directions in different cases, sometimes extending upwards into the pharynx, at others causing the tonsils to project

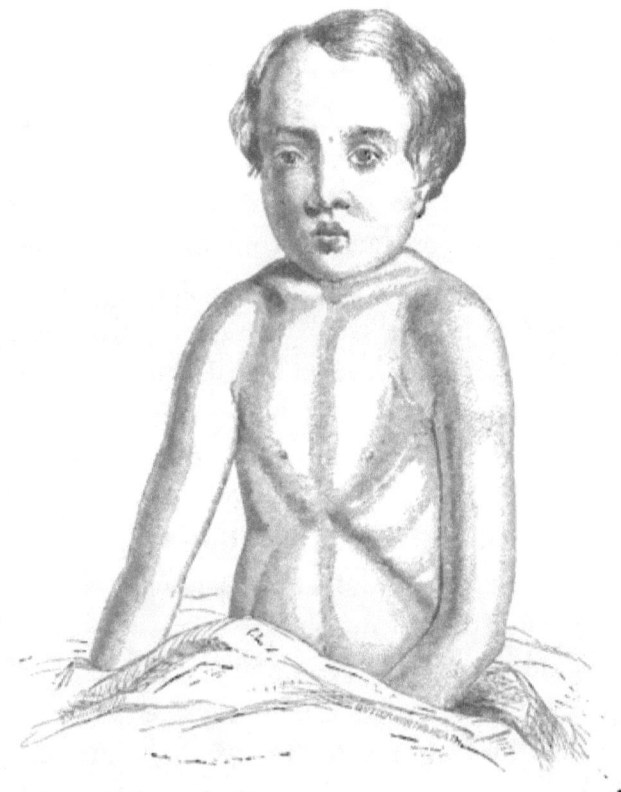

towards the mesian line; and it has been supposed that deafness from obstruction of the eustachian tubes occurs in the former cases and not in the latter. The anatomy of the parts contradicts this supposition; the tonsils are situated considerably lower than the opening of the eustachian tubes, and their enlargement cannot obstruct that orifice. There is evidence also, from the results of treatment, that the deafness is independent of the tonsils. I have removed the tonsils, and the deafness has not been cured; and on the other hand the application of a solution of nitrate of silver to the fauces will sometimes remove the deafness, the tonsils remaining undiminished in size. The deafness in these cases, if connected with the condition of the throat at all (which is very probably the case), is due rather to thickening of the mucous membrane of the eustachian tube or tympanum.

There is also an alteration of the voice in these children which is very characteristic; a thickness and fulness of speech, as if the child's mouth were full; and there is frequently a hacking cough, which may give rise to a suspicion of disease of the lungs.

Upon looking into the throat there is generally seen a quantity of white frothy mucus hanging about the anterior pillars of the fauces, and immediately behind these, on either side of the throat, two large rounded masses, the enlarged tonsils. These vary greatly in size; sometimes there exists a central passage for the food and for the uvula; sometimes the tonsils are so large as nearly to meet; sometimes they are completely in contact, so that it is a mystery how the food gets down the throat, or air into the lungs. The anterior pillar of the fauces often seems flattened out and expanded over them.

In pressing the finger upon these masses, which is

unattended by pain, they are usually felt to be softish and somewhat moveable, and the finger may generally be passed between them and the posterior wall of the pharynx. They may sometimes be seen extending some distance down the throat.

The most various remedies have been recommended for the *Treatment* of this affection. Local applications, as of nitrate of silver, iodine, tannic acid dissolved in spirit, &c., may be of slight service in some cases. I prefer, however, the exhibition of internal remedies —quinine, cod-liver oil, iron. An attempt may be made by a prolonged course of these agents to obtain a diminution of the swelling. I do not think any external applications to the neck are of the slightest service. Change of air, taking the children to the sea-side, with good exercise, and salt-water sponging, I have found of great benefit; and in some cases the surgeon has the satisfaction of finding that the enlargement almost or entirely disappears. If it does not, the question arises what should be done? If the child is not suffering much disturbance in the functions of deglutition or respiration—if, in fact, there is simply the existence of enlarged tonsils, not giving rise to inconvenience, I would leave them alone, having seen that, as the child grows, the tonsils do not seem to increase in a corresponding degree. On the other hand, if the symptoms I have mentioned, and more especially the deformity of the chest, do not become ameliorated, I should perform excision.

Previous to the introduction of the instrument called the tonsil guillotine, shown in the accom-

DISEASES OF THE MOUTH AND TONSILS. 47

panying woodcut, excision of the tonsil was almost impossible in children, owing to their want of self-control. This instrument consists of a metallic

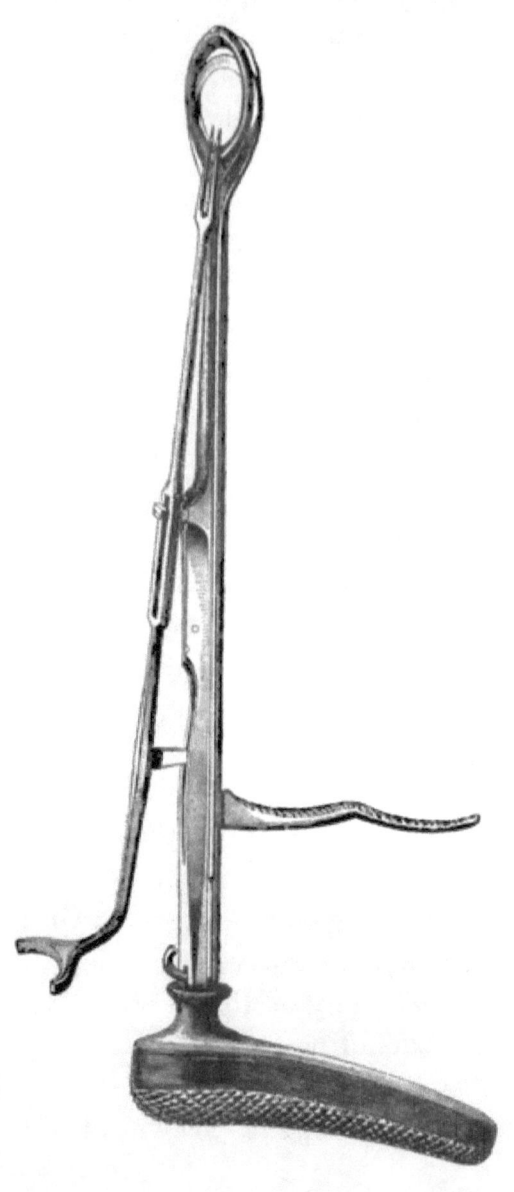

ring fixed at the end of a rod about eight inches in length; this ring is designed to embrace the tonsil, and is divided horizontally so as to conceal within it a moveable cutting blade. Attached to the stem is a two-pronged fork, by which the tonsil is pierced and drawn from its bed, while by the action of a spring the blade above mentioned performs the excision. Without such an instrument it is almost impossible to excise the tonsil in children. Great care, however, should be taken by the surgeon that the blade is sharp, and the instrument in good order. Its disadvantages are, that it is expensive, and after every operation requires careful attention.

In using it, if we are about to remove the left tonsil, the surgeon's left forefinger should depress the tongue, while an assistant holds the head steady, opposite a strong light. The surgeon then, having the instrument in his right hand, should insinuate the tonsil within the ring (and it is in this that his dexterity is to be shown); the tonsil is then to be transfixed by the prong, and the blade drawn towards the operator, by which means the tonsil is removed. The instrument is so made that the blade cannot be used unless the prong be first elevated, and the tonsil thereby drawn out of its bed. In operating on the right tonsil, the surgeon's right forefinger must be used to depress the tongue, and the instrument held in the left hand. It is desirable to remove the whole of the gland; if any part be left, it will be very apt to enlarge again.

Is it necessary to remove both tonsils at the same

sitting, when the disease affects both sides? If the tonsil does not project beyond the mesian line, it is my rule to defer the removal of the second for a fortnight, if circumstances permit. It is less trying to the child, and I have very frequently found that the comfort given by the removal of the one induces the little patient to submit quite willingly the second time. The operation is entirely painless.

When thus performed, I have never seen any ill consequences arise from the attendant hæmorrhage; nor is it greater than that which accompanies other simple operations. Cold water, or in the summer iced water should be used to wash the mouth. No other styptics are desirable. When the forceps and knife are employed there is a risk of wounding the anterior arch of the palate, and if this is done, considerable hæmorrhage may ensue. As to any probability of the carotid artery being wounded, that is quite out of the question.

The after treatment consists in feeding the child on bread and milk, or other liquid diet, as the throat is tender for a few days. Solid food may be given as soon as it can be swallowed.

Having performed this operation for many years and very frequently, I am not aware of any ill result occurring from it, and many of the patients have come under my observation eight or ten years after having submitted to it.

CHAPTER IV.

AFFECTIONS OF THE LARYNX AND TRACHEA.

I. ACCIDENTS TO THE LARYNX.

THE *diseases* of the larynx belong to the physician, but the surgeon's aid is frequently demanded in connexion with accidents to it. A child, for example, is locked in a room by its mother in the winter time, and amuses itself by drinking out of the spout of the kettle; or some diluted acid may be left in a cup, and the child drinks it. In the former case, there is seldom any apparent injury to the mouth, as no child would be likely to drink *boiling* water; in the latter, little white patches will be observed occasionally. The surgeon, when summoned after such an accident, will find the little patient suffering from difficulty of breathing, generally not of a very urgent character, but the child is restless and uneasy, and there is a peculiar hoarse or crowing noise attending its respiration. The countenance looks full, and the veins of the face and neck are more or less turgid. The period at which these symptoms arise, and their degree of urgency, vary very much, according to the temperature of the water, or strength of the irritant. In some cases they commence in a few minutes, in others, some hours elapse. The symptoms are due to œdema of the glottis, and rapidly increase in severity. There is present the peculiar sinking in of the neck, just above the sternum, and of the abdomen below the ensiform cartilage, which always characterizes obstruction of the glottis in children.

The great question in these cases is whether or not tracheotomy should be performed. At Guy's Hospital the usual practice is to perform it when the danger of suffocation seems imminent. But the results are, upon the whole, unfavourable, owing to extension of disease into the lung; and I believe they ever will be so in all large hospitals, unless a special small ward, carefully regulated as to temperature and supply of moisture, be provided for such cases. On the other hand, the results are often unfavourable if tracheotomy is not had recourse to.

Case 1.—Swallowing hot water: tracheotomy: death. —J. C., aged eighteen months, admitted into Guy's Hospital, May, 1858: the evening before admission, drank from the tea-kettle. Difficulty of breathing came on in the night; the following morning the child was brought to the hospital. Being apparently moribund, tracheotomy was performed, under chloroform, with instant relief. The patient progressed favourably for two days, and then difficulty of breathing supervened, and the child soon died. *Postmortem examination.*—Pneumonia was found to have been the cause of death. The glottis and epiglottis were swollen from sub-mucous infiltration, but not to any great extent. The air-tubes were inflamed and filled with secretion. The trachea was most inflamed around the opening. The lower lobes of both lungs were consolidated throughout.

Case 2.—*Swallowing hot water: symptoms not urgent: death half an hour after admission.*—E. L. H., aged three and a half years, admitted into Guy's Hospital, January, 1855, a stout well-nourished child; had drank out of the tea-kettle about five o'clock in the afternoon. Was brought to the hospital at ten. (It was mild weather.) There was a little difficulty of breathing, but no urgent symptoms. The child died suddenly half an hour after admission. *Post-mortem examination.*—On the mucous membrane of the pharynx and posterior pillars of the fauces there were several shreds, apparently from ruptured vesicles; also on the laryngeal surface of the epiglottis. The epiglottis itself was œdematous and thickened.

The upper part of the larynx, as far as the vocal chords, was white in colour and filled with mucus. The bronchi were congested and full of mucus.

Case 3.—In the following instance, tracheotomy was performed successfully :—E. W., aged three years, was admitted into Guy's Hospital on February 7th, 1854. One hour before admission he had drunk some water out of a tea-kettle. There was much embarrassment of respiration. Tracheotomy was performed. Immediate relief was given by the operation; but the next day there was much constitutional disturbance. This, however, subsided. The tube was removed at the end of a week, and the boy went out well on the 1st of March.

Case 4.—The following is an example of recovery

under treatment by antimony:—C. B., aged two years, was admitted into Guy's Hospital at eleven p.m. He had swallowed water from a tea-kettle about five o'clock. Difficulty of breathing was first noticed at ten. When admitted, the inspiration was accompanied by the usual crowing noise of obstructed larynx; and the difficulty of breathing was apparently increasing, though it was not as yet sufficient to cause lividity of the face. Tracheotomy was proposed, but it was resolved that the operation should be deferred until more urgent symptoms arose, and meanwhile to give antimony, so as to produce nausea. Eighteen minims of antimonial wine were given at once, and repeated in an hour; the dose was then continued every second hour for forty-eight hours. He was sick after the first dose, and again in the course of the day. He began to improve at once, and the third day was well.

These are examples which I have taken indiscriminately from the notes of a large number similar to them. And they will indicate the difficulty which cases of this sort present to the judgment of the surgeon. My own impression is unfavourable to the operation of tracheotomy. If it diminishes the risk from one source of danger, it introduces others which did not previously exist, and it is hardly possible to say, until life is all but extinct, whether it will be necessary or not. I should, at any rate, be disposed to defer the operation until the very last moment; until one might almost say the patient had ceased to breathe. The chance of recovery by opening the trachea is not even then hopeless, as a case which I shall mention, when discussing the question of tracheotomy in croup, will show, and, at least, no risk is incurred of preventing a recovery that might otherwise have taken place.

What, then, should be done? We must not expect to save all. No one can remember the delicacy of the membrane lining the air-passages without being surprised that such cases should ever recover. Of all the means that can be employed, I think antimony and ipecacuanha the most effectual. They must be given in full doses, varying according to the age of the child, frequently repeated, perhaps every quarter of an hour at first, gradually increasing the intervals; and they should be continued until there is decided relief to the symptoms. I have never known them, in these cases, produce vomiting. It is my impression that, where a remedy is required, there is a marked tolerance of it. I have given to a child fourteen months old, under these circumstances, half a drachm of antimonial wine every half hour, for six hours, and the child recovered without any ill effects. I am happy to find that Dr. Bevan, of Dublin, recommends a somewhat similar plan of treatment.

If the patient continues to get worse, tracheotomy can be had recourse to. I am not prepared to say that if the child could be placed in a warm room, and the temperature equably maintained at about 80°, the air being fully saturated with moisture, in fact in the atmosphere of an orchid house, tracheotomy might not be advisably had recourse to at once. But practically, this is attended, except in a few instances, with very great difficulty.

II. FOREIGN BODIES IN THE AIR PASSAGES.

There are no accidents in which an accurate diagnosis is more important than those of the lodgment of foreign bodies in the trachea or larynx, and none which demand a more patient and careful investigation. The greatest certainty may, almost always, be attained by sufficient care; but there are scarcely any cases which are more deceptive on a superficial examination, or in which the history is more likely to mislead. In all the patients, the symptoms are urgent at first; sometimes these urgent symptoms continue until the surgeon sees the patient; sometimes they rapidly abate, and are succeeded by others much less severe.

These differences are due to various causes; the size of the foreign body, and its position, whether in the glottis, free in the trachea, or impacted in the right bronchus. If it be in the glottis, or freely moving in the trachea, the symptoms are very much alike; whereas, if it be lodged in the right bronchus, the absence of air in the right lung renders the diagnosis tolerably easy. I say, impacted in the *right* bronchus, for, according to my experience, these bodies pass into the right side. The greater size, the shortness, and the direction of the right bronchus, together with the circumstance that the point of bifurcation of the trachea is on a plane to the left of the mesian line of the body, account for this fact. It is a curious circumstance, that so far as my experience extends, children with foreign bodies in the air passages, if they lie down at all, invariably lie on their right side. If this symptom were confined to the cases in which the foreign body has fallen into the right bronchus, it might be explained by a desire to diminish the pressure; but I have observed it in all cases, wherever the foreign body may be situated.

The accident takes place during a deep inspiration, and as soon as ever the foreign body reaches the air passages, violent coughing ensues, and fortunate is the patient if it be thus expelled. More frequently, however, it has already passed through or into the glottis, and the narrowness of the passage prevents its ejection; there seems to be a greater difficulty in the passage of a body outwards through the glottis, than inwards, probably in part from a spasmodic action excited by its presence. Sometimes intense dyspnœa, laboured respiration, incessant attempts at coughing, and rattling of mucus in the air tubes succeed and are accompanied with great distress; the countenance being anxious, the pulse feeble, and the extremities cold. The patient, indeed, appears to be dying. Upon examining the chest, we find more or less diminution of the respiratory murmur, sometimes confined to the right side. These symptoms, with the exception of the stethoscopic indications, may, and most frequently do, all pass off, and the child appears pretty well, though it is still disturbed and uncomfortable, and will sometimes even tell us that there is something moving up and down in its throat; and the ear applied, by means of the stethoscope, will confirm the fact. The mitigation of the symptoms is not permanent; and the severer ones recur at various intervals, unless the foreign body becomes impacted in the bronchi; in which case a different train of symptoms occurs; those, namely, of bronchitis or pneumonia, followed, perhaps, by those of phthisis. There seems to be good ground, indeed, for the opinion, that the permanent presence of a foreign body in the lung is sure, at some time or other, to prove fatal.

When we have discovered the existence of a foreign body in the air passages, what treatment should be employed? Should we, or should we not, have recourse to the plan of inverting the child for an instant, and shaking him, with a view of dislodging the body, and suffering it to escape by its own weight? I should, myself, be very cautious in doing this, and would never venture upon it without being ready immediately to perform tracheotomy. There is a risk of instant death from the lodgment of the body in the rima of the glottis; a piece of money, for example, might fall flat upon the glottis and completely close that opening. There is, indeed, but one safe plan of treatment, and that is to open the trachea, and remove the foreign body.

Case 1.—I. M., aged five, a sickly, ill-conditioned child, living in Bermondsey, was admitted into Guy's Hospital, 11th August, 1855, at four p.m. The history given by the mother went to show that the child was whistling through a plum-stone ground for the purpose, and that as it annoyed her, and he would not desist, she suddenly slapped him on the back. This caused him to take a deep inspiration, and with the rush of air into the larynx, the stone must have entered also, as he was immediately seized with dyspnœa and attempted to cough; and so urgent were the symptoms that, thinking the child to be dying, she immediately brought him to the hospital. I saw him within a few minutes after admission, and found him suffering from intense dyspnœa, laboured respiration, incessant attempts at coughing with rattling of mucus in the air-tubes, and great distress; the countenance was livid, the extremities were cold, and the pulse feeble; he was lying on his right side and could not be moved without the cough becoming aggravated; when quiet there was no respiratory murmur on that side of the chest. The little patient apparently was dying. I felt that time was not to be lost, that the history and symptoms clearly pointed to a foreign body in the air-passages, probably in the right bronchus, and that the performance of the operation of tracheotomy offered the only available means of relief. Having previously satisfied myself by the passage of the probang that no substance was fixed in the œsophagus or pharynx, I cut down upon the trachea in the usual manner; a little venous hæmorrhage occurred, and without waiting to fix the tube with a tenaculum, I at once divided four of its rings. No sooner was this done than with the violent expiratory efforts always observed when the trachea is first opened, the stone, surrounded with mucus, was forcibly ejected, with immediate relief to all the urgent symptoms; the placid look of the child's countenance clearly indicating the comfort he experienced from the removal of the body. The respiration quickly became natural. The little patient was placed in the warmest bed in the ward, and a piece of wet lint applied over the wound. A slight attack of bronchitis followed, which subsided without any treatment, and he was discharged on the 26th August, fifteen days after the operation, perfectly recovered, the wound in the neck having quite healed.

Case 2.—T. N. P., aged nine, a delicate-looking, but sharp and intelligent boy, living in Hackney Fields, was admitted into Guy's Hospital, 9th July, 1856. The history given was, that five days previously he was playing with some cherry-stones, and placed one in his mouth, that it disappeared from that cavity, and, as the boy thought, "went the wrong way." He was immediately seized with difficulty of breathing, fell down, and was unable to walk home, but upon arriving there soon became much relieved. During the five days previous to admission, he was frequently seized with dyspnœa and became almost suffocated; then the symptoms as suddenly abated, and he would play about quite comfortably; he was unable to sleep at night in any other positions than sitting upright or lying on his right side. He ate and drank well, and with the exception of the attacks of dyspnœa, which occurred principally in the evening, would not have been thought to have had anything ailing him. He was taken to several practitioners and to two hospitals, but the nature of his complaint was not diagnosed until he was seen by Mr. Wallace, of the Hackney Road, who sent him to the hospital. I saw the patient soon after admission, and found him in one of his usual evening attacks of dyspnœa, he was almost suffocated, and complaining that something was moving up and down his throat; he was lying on his right side, and could not turn over to the left, nor lie in any other position without feeling the most intense choking sensation; in a few minutes this passed off, and he appeared quite comfortable. The respiratory murmur was heard in both lungs. The history and symptoms clearly indicated that there was a foreign substance in the air-tube, and as it appeared that I should gain nothing by delay, I performed the operation of tracheotomy in the usual manner; there was a good deal of troublesome hæmorrhage entirely of a venous character, which did not prevent my opening the trachea with a knife immediately I had exposed it. A violent expiratory effort discharged the cherry-stone, and he had no return of his dyspnœa. The same treatment was adopted as in the other case, and he was discharged from the hospital 21st July, quite well, twelve days after the operation. He was re-admitted on the 26th, having caught a slight cold, and there being a little erysipelatous blush about the neck, which quickly subsided, and he was again discharged on the 5th August.

Case 3.—F. H., aged ten months, a pale, delicate little child, residing at Lambeth, was admitted with its mother into Guy's Hospital at half-past eight, p.m., 28th February, 1857. The mother stated that two hours before admission the child was playing on the floor with a small marble and a piece of crust, that it appeared suddenly to be choking, and she missed both the substances, and as she could not find them, concluded that the child had put them into its mouth; she therefore thrust her finger down its throat, but nothing could be brought away. The severe symptoms subsided, and though the infant was evidently not quite comfortable, there appeared to be but little ailing it until an attack of dyspnœa occurred; the child was then taken to a neighbouring practitioner, who sent it to the hospital. Upon admission, the little patient appeared to be *comparatively* comfortable when lying in its mother's arms on its right side; immediately she placed it on its left to take the breast dyspnœa occurred, and the child gasped for breath. The breathing was stridulous and croupy, and evidently of such a character as pointed to the existence of a foreign body in the air passages, and demanded immediate interference; the patient was restless and crying, which prevented an examination of the lungs. Having satisfied myself, by the passage of a probang, that the pharynx and œsophagus were clear, I proceeded to perform the operation of tracheotomy, which, upon a child so young, presented some difficulty, though I must confess I had anticipated more than I found. The venous hæmorrhage was the only trouble and was very considerable; being unable to arrest it, I at once plunged the knife into the trachea and divided several rings, cutting from below upward; the bleeding quickly ceased upon the opening being made, but from the amount of blood which had been lost the child became faint, the lips pale, and the breathing scarcely to be noticed; artificial respiration was resorted to in the manner recommended by Dr. Marshall Hall, viz., by rolling the child on its side and back again, and it quickly recovered. There was no particular relief to the child when the trachea was opened, nothing being ejected as in the other cases. I tried to feel with the tip of my little finger if there was anything above the artificial opening, but the trachea was too small to admit it, and expecting from the history given by the mother to find a marble or piece of crust of bread come out, I did not of course pass a trachea tube, rather waiting in expectation that a violent fit of coughing would expel one of the two substances said to have been lost, but I was disappointed. The child struggled to get on its hands and knees, and we allowed it to do so, when it immediately rolled over to the right side, that apparently being the most comfortable position.

The little sufferer was now put to bed, but was extremely exhausted from loss of blood, and apparently sinking; indeed, it never rallied from the effects of the operation, but died in four hours, without having taken the breast or exhibited signs of perfect recovery from the collapse.

The post-mortem examination was made fourteen hours after death.

The lungs were found healthy and collapsed; the bronchial tubes were quite clear; no blood had passed into them from the wound. Upon removing the larynx, trachea, and œsophagus, with parts about the neck, a thin piece of bone was seen just emerging from the upper part of the larynx, and lying in the rima glottidis. Upon laying open the larynx, the piece of bone was found fixed in the rima, extending above the superior and below the inferior vocal chords, with several irregular points sticking firmly into the mucous membrane anteriorly and posteriorly, and apparently completely preventing its passage either up or down. The position of the bone is shown in the accompanying woodcut. The opening I had made in the trachea was through five of its rings and in the median line. There was not any œdema about the parts. Upon investigation, it appeared that most probably this foreign body was a piece of mutton

bone. The next day, upon questioning the mother, she stated that she had given the child a small quantity of broth, previous to placing it upon the ground to play; but declared most positively that there were not any symptoms for a quarter of an hour at least after the broth had been taken. The marble and piece of bread were afterwards found upon the floor.

When the foreign body, passing into the trachea, falls at once into the right bronchus, a peculiar class of symptoms occurs, as shown in the following case.

Early in the morning of June 6th, 1859, a little boy, aged three years, was brought to Guy's Hospital, with the following history. At two o'clock on the day before, while playing with a French bean he put it into his mouth, and soon afterwards suddenly became choked. He coughed, struggled violently, and became livid in the face. His father snatched him up and hurried with him to the nearest surgeon. On his way the child suddenly became relieved. Some medicine was ordered, and the danger being considered over, he was taken home again. Gradually during the evening difficulty of breathing developed itself, and he passed a restless night. Towards morning the dyspnœa and restlessness became aggravated, and at eight a.m. the boy was brought to the hospital. I saw the patient very soon after his admission, and found him with livid lips, gasping for breath, and showing all the signs of urgent thoracic distress. On auscultation no respiratory murmur could be heard on the right side of the chest, the right lung appearing quite impervious to air. With the exception of the first paroxysm there had been no cough throughout. I felt satisfied from the history of the case, and from the symptoms, that the bean was lodged in the right bronchus. The dyspnœa was fast increasing. The veins of the neck were already turgid and prominent, and unless relief could be afforded it was clear the child must sink. Under these circumstances I at

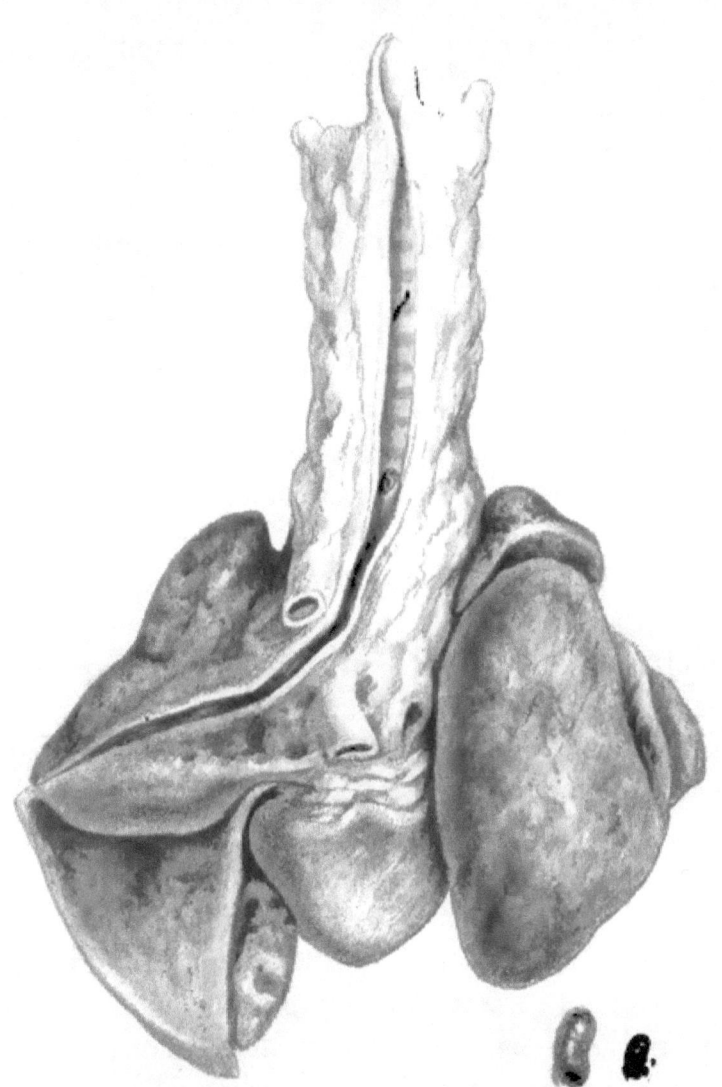

once performed tracheotomy. Chloroform was not given, owing to the embarrassment of the circulation; the neck was unusually thick and fat; the veins which were divided in the incisions bled most profusely. On exposing the trachea I made the slit in it much longer than usual, with the hope that the bean would be expelled. The child's dyspnœa appeared to be greatly relieved by the loss of blood, and when the operation was completed he seemed to be much more comfortable than before. Attempts were made to excite cough in the hope that the bean would be expelled, but these failed. The child was now laid on its face and struck sharply on the back, and subsequently he was also held with the head downwards, the strokes on the back being repeated in that position. But all these efforts to effect the dislodgment of the bean proved fruitless, and were at length of necessity discontinued. On the day following the operation the child was feverish. I attempted by the introduction of a pair of long curved forceps, which had been procured for the purpose, to seize the bean, but could not succeed in doing so. And I may here remark that I have never seen the use of such instruments of any avail. The following day, at six in the evening, the child, who had been gradually getting worse, died. At the autopsy the bean was found firmly plugged in the right bronchus. It closed the orifice of that bronchus much as a cork might have done, and projected into the trachea above. The right lung was collapsed and solid, but had no traces of inflammation. The bean had swollen from the moisture of the part; at first it had probably acted the part of a ball valve to the bronchus, and while allowing the air to escape upwards, prevented its ingress; the swelling had no doubt caused it to close the orifice so firmly. The smooth rounded extremity which projected into the trachea could not have been effectually seized by the forceps. The accompanying lithograph represents a reduced view of the parts, seen from behind; the trachea and left bronchus being slit up. The slit-like opening in the upper part of the trachea represents the wound made in the operation. Below it, will be seen a small portion of the bean projecting from the right bronchus. The bean also is represented, relatively reduced in size; the larger of the two figures indicating the amount of swelling which it had undergone within the lung.

III. THE OPERATION OF TRACHEOTOMY.

Having discussed the circumstances under which tracheotomy should be performed, an operation which I prefer, and which I believe is now by universal consent preferred to laryngotomy, I shall proceed to describe the steps of the operation, which, in these young children, is a difficult and disagreeable one. The patient should be placed under the influence of chloroform. This has been objected to by some, but the late Dr. Snow corroborated my opinion, that there was no peculiar risk attending its use in these cases, and the comfort to the operator is immense. The little patient should lie upon a pillow, with its head thrown as far back as possible, while the surgeon carefully inspects the neck, and makes up his mind in respect to the various stages of the operation. It is one which requires especial coolness and decision, and should be undertaken with full deliberation. Occasionally tracheotomy must be performed with the greatest rapidity, in order to be of any avail, but it is necessary to avoid a nervous haste. I suppose it must have been from want of due deliberation that the pleura has been punctured in the operation, and the trocar even passed into the spinal column! An incision should be made in the mesian line, extending from the thyroid cartilage to half an inch above the sternum. It should pass through the skin and fascia, and will most probably divide the anterior jugular vein, both ends of which may be tied, if the hæmorrhage from it causes any embarrassment to the operator. The muscles attached to the sternum below, and to the os hyoides and thyroid cartilage above, now come into view. Those on the right side should be separated from their fellows on the left, if possible, but there need be no hesitation in cutting through their fibres. A couple of retractors should be placed in the wound, to hold the muscles on one side. A great deal of venous hæmorrhage will now most probably occur. If any arteries are wounded, they must be tied, and the veins also, if the bleeding from them is very troublesome. The lower border of the isthmus of the thyroid body is brought into view, and a dense fascia covering the trachea, which must be divided, and with it, now and then, the middle thyroid artery, which gives rise to very smart hæmorrhage. A knife should next be carefully passed into the rings of the trachea, at the lower border of the wound, and an incision made directly upwards, dividing two or three of them. In doing this, it is necessary to avoid wounding the posterior surface of the trachea. This is an accident not unlikely to occur, owing to the yielding of the anterior wall of the tube, and when it does, it gives rise to a disagreeable hæmorrhage. It is advisable, as a rule, to fix the trachea by means of a hook introduced into it previous to making this incision. The trachea being opened, a violent fit of coughing immediately takes place, and mucus and blood are forcibly ejected.

If the operation is performed on account of œdema of the glottis, it is advisable to pass a canula, armed with a blunt trocar, through this opening, and to fix it there by means of tapes tied at the back of the neck. When the trocar and canula are first introduced into the trachea, the most violent expiratory efforts ensue, and one would imagine that immediate death was about to take place, but these symptoms pass off in less than a minute, and the breathing becomes quite natural through the tube.

If the operation is performed on account of the presence of a foreign body in the passage, the opening should, of course, be left free. The foreign body is frequently expelled at the first effort; and if not, the little finger, or a probe, may be passed upwards to examine the region of the glottis, and a pair of long forceps, downwards, to explore the bronchi. If nothing is found, the child must be left, trusting that upon a fit of coughing ensuing, the substance may be expelled.

It is generally recommended that the trachea should not be opened until all external hæmorrhage has ceased; and the propriety of this rule, as regards a bleeding artery, no one can doubt. But when the hæmorrhage is altogether venous, as it usually is in these cases, the most advisable plan appears to me to be, to relieve as quickly as possible the embarrassed circulation, and thus allow the blood to return to the heart. I do not hesitate, therefore, especially when the operation is performed to avert impending suffocation, to pass the knife into the trachea as soon as possible, and I then turn the child quickly over on its face. Almost invariably, immediately the opening is made, and respiration has become established by the wound, the congestion of the veins is relieved, and the hæmorrhage ceases.

The after treatment of the wound must depend upon the nature of the case, and the results which have been attained. Supposing the operation to have been performed for the removal of a foreign body, if the latter has been immediately ejected, it will be advisable to close the opening in the skin as quickly as possible; in fact, to treat it as a simple incised wound, and apply the interrupted suture. If the foreign body have not been expelled, so that the wound cannot be closed, a piece of lint dipped in water should be laid over the incision.

Lastly, if the operation have been performed for obstruction of the glottis, and the canula therefore is left in, it should be covered with a piece of fine gauze. In the last two cases, it is advisable to bring together the edges of the wound at the upper and lower parts, and to unite them by sutures, as far as can be done. And the patient should be placed in a warm moist atmosphere. When the canula is left in the wound, it is necessary for some one to sit by the bed, to see that the passage does not become obstructed by thick mucus, to which there is a great liability, as owing to the irritation caused by the tube, the mucous membrane of the trachea pours out a greatly increased secretion. Children have died from neglect of this precaution.

It is very probable that if the operation have been performed for œdema of the glottis, the obstruction may have passed off in the course of two or three days, in which case the tube may be removed, and it is seldom necessary to keep it in longer than a week. For the purpose of cleaning the canula, and preventing the accumulation of mucus, it is usual to use a second tube fitting within it, which may be removed as often as is necessary.

I have noticed a curious circumstance in connexion with the operation of tracheotomy. In several cases in which it has been performed for œdema of the glottis, on or about the second day after the operation, milk, or whatever liquid food the child may be taking will make its way in small quantities through the wound. The natural supposition at first is, of course, that the œsophagus has been punctured, but this is not the case. By the fourth or fifth day it entirely ceases. I presume that it is owing to the swollen larynx and epiglottis not performing their functions properly. The epiglottis probably does not close upon the larynx, and the œdematous condition of the latter depriving it of its natural irritability, the fluid passes over its surface without exciting cough.

The surgeon may be called upon to perform tracheotomy on children for other causes than those already mentioned; for instance, the pressure of enlarged glands, or croup. The former case is extremely rare, and when speaking hereafter of enlarged glands in the neck, I shall have to relate a case in which these glands were extraordinarily increased in size, and yet no pressure on the air passages resulted. The following case, however, occurred in Guy's Hospital, in the early part of this year, and was operated on by Mr. Bryant.

"G. W., aged ten, was admitted January 19th, 1860. He was a strumous-looking lad, who for some months had been the subject of abscesses in the neck. Two weeks previously he had first experienced a difficulty in breathing, and discovered a swelling over the trachea; this swelling gradually increased, and for five days before his admission his respiration had become more difficult. When seen, suffocation appeared imminent; his face and neck were much congested; head thrown back, respiration laboured and almost stridulous. Upon examining his trachea, a tumour was detected growing over the sternum, and passing upwards to the thyroid cartilage. It was firm and immovable, and had evidently, by its pressure, obstructed the passage of the trachea; the cricoid cartilage, and the trachea, were not to be felt.

The tumour appeared to be connected with the thyroid gland, involving its isthmus."* As death seemed imminent, tracheotomy was performed. The swelling gradually diminished, and the boy went out well. Under the head of accidents to the œsophagus will be found a case in which tracheotomy was performed on account of the swallowing of strong sulphuric acid.

Much discussion has arisen lately respecting the performance of tracheotomy in cases of croup. It is not for me to enter into this controversy. I have performed the operation in various periods of the disease, and with varied success. In any early stage I have doubted, and still doubt, whether it should be performed at all; it appears to me only to add increased risk to a fearful disease. In the more advanced stages of the disease, when the patient seems about to succumb, the surgeon need not fear bringing an operation into disrepute by plunging a knife into the trachea. Nay, I doubt much whether when a patient is moribund be not the most advisable time for the proceeding. In the following case, for the report of which I am indebted to my friend Dr. Wilks, it will be seen that the patient was not operated upon until quite *in articulo mortis*, yet the result was most satisfactory.

Case. — *Laryngitis; Tracheotomy; Recovery.* — John P., aged four years, first seen by Dr. Wilks on July 8th, 1857. The child was then suffering from laryngitis, the trachea and bronchi being only slightly

---

* *Medical Times and Gazette, April 21st,* 1860.

affected. There was much fever, loud croupy respiration, voice whispering and hoarse. The symptoms were not urgent, nor was danger apprehended, except from the known importance of the complaint. He was said to have been ill only two days, but afterwards it was learned that he had been hoarse and husky for two or three weeks past. An emetic was given which afforded much relief; it was followed by a mixture of ipecacuanha and antimony, and by calomel powders.

July 9 (the following day), the child was much the same; no very urgent symptoms, although the laryngeal obstruction was still considerable, and the breathing croupy. Continued the medicines.

July 10.—The child was not seen in the day. In the evening the father stated that the patient had been gradually growing worse, and at that time he believed him to be dying, the difficulty of breathing having rapidly increased. I was now requested to accompany Dr. Wilks, and to perform tracheotomy, if an operation seemed advisable. We found the child actually dying, and probably within a few minutes of his dissolution. He was lying upon his side perfectly insensible, face and lips quite livid, faint efforts of respiration going on, but the air scarcely entering the chest, the pulse imperceptible, and the skin cold. It was evident that not a moment was to be lost, and therefore, I immediately opened the trachea by a longitudinal incision with the scalpel, and thrust in the trocar. The effect was instantaneous, and (although not a novel sight) was truly surprising. The air rushed in, and immediately the most violent respiratory efforts were made; the inspirations filled the chest, and at each expiration large quantities of viscid mucus were ejected to a great distance from the patient. Up to this time perfect insensibility existed, but now the lividity faded away from the face, and the lips resumed their natural pink hue; the boy opened his eyes, and recognised those around him; in fact, he was restored to life. He was soon after sick (the emetic taken some hours before not having operated), the skin became warm, he swallowed some milk, and in a quarter of an hour after the operation he was in a more comfortable condition than he had been in for several days.

July 11.—Very comfortable; breathing tranquil; much mucus discharged through canula.

July 12.—Improving; the febrile symptoms much less, and the tongue cleaning. Scarcely any râles in the chest. The tube has been several times almost blocked up by very tenacious mucus, requiring removal.

July 13.—Nearly all the mucous discharge has ceased, there being now only a little purulent matter oozing from the wound. Upon closing the opening of the canula with the finger, respiration is entirely stopped, as no air yet penetrates the larynx. As, however, there had been for some hours a constant hacking cough, it was thought advisable to remove the canula. This was done with much comfort to the patient, the wound remaining sufficiently patent to allow the ingress of air.

July 15.—The child had breathed very well through

the opening; all febrile symptoms have disappeared. Since the operation he has taken small doses of the original medicine, ipecacuanha and antimony. Diet, milk only.

July 17.—Convalescent. Wound slowly healing, and passage contracting, the air now entering, in part, through the natural passage.

July 19.—Breathes almost entirely by the natural way. Wound filled with granulations, and only a slight whiff of air to be detected escaping from it. Speaks in a faint whisper for the first time.

July 22.—Child up and dressed. Breathes entirely in the natural way; voice whispering and hoarse; wound apparently closed, but when he drinks he frequently chokes, and then a little of the fluid is seen oozing through the wound. During the next fortnight, although the laryngeal symptoms were disappearing, and the case so far doing well, the patient had two severe febrile attacks, which retarded his cure. These were apparently gastric, and were probably due to the resumption of his ordinary food.

Aug. 6.—The febrile symptoms abating; ordered quinine and cod-liver oil, when he again began rapidly to improve; his appetite returned; he gained flesh; he was able to go out of doors, and at the end of the month he was perfectly well.

Within the last fortnight I have seen this little fellow; he is well and hearty.

CHAPTER V.

AFFECTIONS OF THE PHARYNX AND ŒSOPHAGUS.

I. DISEASES AND INJURIES OF THE PHARYNX AND ŒSOPHAGUS.

EVEN in the adult, the pharynx and œsophagus are subject to few diseases except those arising from accident. And the following are the only ones which I am aware of that demand our attention in the child.

1. We now and then meet with *follicular growths* from the posterior walls of the pharynx, pedunculated in character, and resembling small polypi. I have seen them in children, about the age of twelve, though rarely. One young lady was brought to me, at that age, who had these growths at the back of the pharynx, one on each side. I was at first doubtful, from their position, whether they were not attached to the tonsil. I removed them with a probe-pointed bistoury, and they have not returned.

2. An abscess may form behind the posterior wall of the pharynx, constituting what is known as *post pharyngeal abscess*. This may arise from disease of the cervical vertebræ, and I have known it occur from the irritation of an impacted fish bone. It is very rare. In the cases I have seen in children, it has commenced without pain, and the first symptom complained of

was difficulty of swallowing followed by impeded respiration. The true nature of these cases is very apt to be mistaken unless a careful examination be made. On opening the mouth widely, and passing the finger to the back of the pharynx, a swelling is discovered, which is soft and fluctuating. The *Treatment* consists in opening the abscess if it impede deglutition or respiration, as its tendency, in consequence of the loose cellular tissue at the back of the pharynx, is to pass upwards or downwards rather than forwards. The *prognosis* of these cases is very uncertain, and depends upon the presence or absence of disease of the vertebræ. The diagnosis and treatment of the latter affection will be discussed in its proper place.

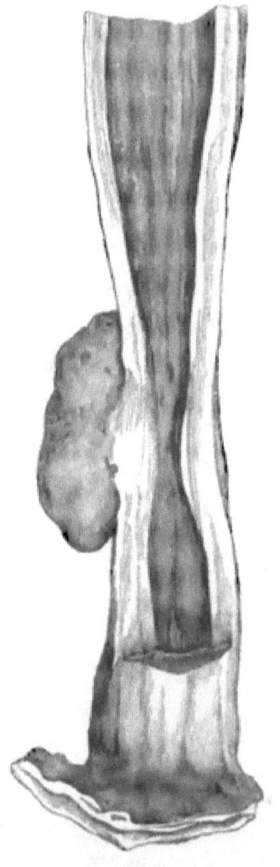

3. The accompanying woodcut represents another class of cases; it is the œsophagus of a boy who had difficulty of swallowing from his infancy, to relieve which a bougie was occasionally passed. He died from fever. It will be seen that, about two inches above the cardiac extremity of the stomach, there is an enlarged gland adherent to the walls of the œsophagus, and pressing upon that tube. There may

be a question whether this was the cause of the thickening of the parietes of the canal, the mucous membrane of which was contracted, and apparently cicatrized for the space of an inch and a half.

II. FOREIGN BODIES IN THE PHARYNX, ETC.

We sometimes find that a child has "bolted" a lump of food, generally of pudding; or it may swallow, by accident or design, various other solid substances; these may lodge at the lower part of the pharynx, or in some part of the œsophagus, and give rise to the most serious symptoms of suffocation. Death by asphyxia has actually resulted from such an accident. There is always a question in these cases, whether the mass is lodged in the œsophagus or in the trachea; and thus arises an universal rule of practice; viz., always to employ the probang before opening the trachea, in cases in which a child is suddenly attacked with symptoms of suffocation, and in which there is no clear history of its having swallowed hot water, or a corrosive liquid. The neglect of this rule I once saw attended with fatal results. A child aged a year and a half, had secretly taken a piece of pudding. It was suddenly seized with difficulty of breathing; the surgeon saw it shortly afterwards, and the trachea was opened; of course with no result. Upon making the post-mortem examination, a large piece of pudding was found in the œsophagus, pressing on and closing the trachea.

The following case of a shilling impacted in the pharynx, was reported to me by the dresser under whose care the child came. L. W., aged seven years, admitted into Guy's Hospital in April, 1858. While playing with some money she had placed a shilling in her mouth; this fell back into the pharynx, and induced symptoms of suffocation. She was immediately brought to the hospital; on looking into the pharynx, the shilling was seen lodged on the left side, and the dresser attempted to seize it with the forceps. In this attempt it slipped down into the stomach. The fæces were watched, but not any trace of the shilling was found.

The *Treatment* of these cases consists in passing the finger down the pharynx as far as possible, and hooking up the foreign body with the nail, if it can be reached. By this means, also, violent efforts at vomiting are caused, which are of course desirable; and the surgeon must boldly persist in forcing the finger as far down the throat as possible, in spite of the cries and struggles of the patient. If not successful in thus dislodging the foreign body, a probang must be passed down the œsophagus. I have never known a case which has not been relieved by one or other of these means. But the probang must be used with a certain amount of caution; the end of that instrument has been seen in the posterior mediastinum in a case of this kind, from having been forced through the walls of the œsophagus.

If a foreign body, not easily soluble in the gastric juice, such as a marble, coin, pin, fishbone, &c., has passed into the stomach, the usual practice is to give an aperient. I think, however, this is unadvisable for the following reason. The effect of an aperient is to render the evacuations from the bowels liquid, thus doing away with the natural security which the semi-solid condition of the contents of the colon affords against the lodgment of a foreign body in any of its numerous sacculi. It is better that the substance whose ejection we desire should be left to the natural action of the intestine, and to take a fixed position among the rest of its contents. When aperients are given it is sometimes necessary to pass the finger up the rectum to dislodge angular substances, such as fishbones or pins.

Irritating liquids, as corrosive poisons, strong acids, caustic alkalies, &c., are sometimes swallowed by children, and injure the œsophagus and stomach.

The following is an illustrative case:—A female child, aged seventeen months, was brought to Guy's Hospital at ten p.m., on the 28th of July, 1857. It was stated by the mother that, at nine in the morning, she gave the child some sulphuric acid by mistake for syrup of buckthorn. She quickly found out her error, and took the child to a neighbouring chemist, who gave her large quantities of magnesia as an antidote. Three hours before admission, great difficulty of breathing occurred, which was rapidly increasing. The child, when taken in, had the usual croupy respiration, and was fighting for breath. I did not hesitate, knowing the history of this case, to perform tracheotomy; it was accomplished without much difficulty, and with scarcely any bleeding; and the child was immensely relieved. She lived until the following day, breathing comfortably. On the post-mortem examination there was found extensive charring of the œsophagus and stomach, great thickening of the epiglottis, and complete obstruction of the larynx.

It may be questioned whether tracheotomy should be performed in cases such as these, where there is a doubt respecting the amount of injury inflicted on the œsophagus and stomach by the irritant. From the relief afforded to this child, and the comfort of its few last hours, I should have no hesitation in again acting in the same way, even in a case which might seem to be hopeless.

For the most part, accidents of this nature are rapidly fatal. If they are not so, other symptoms of injury to the œsophagus, of a permanent character, make their appearance subsequently. I cannot better illustrate this subject than by the following case, which I published in the Guy's Hospital Reports for 1859. It will be observed that the first effects of the irritant having passed off, no further symptoms occurred for a period of fifteen weeks.

Case.—Contraction of Œsophagus from corrosive poison; Gastrotomy.—James G., aged four years and four months, was admitted into Martha Ward on the 2nd of February, 1859, under Dr. Addison's care, in an extremely thin and emaciated condition. I am indebted to Mr. Fagge for the following history, obtained from the parents. Seventeen weeks before admission the child swallowed some corrosive poison, supposed to be a solution of potash, or caustic alkali, used for bleaching and cleaning linen; he was instantly seized with violent vomiting, bringing up frothy mucus, and either at the time, or shortly afterwards, the vomited matter contained two or three teaspoonfuls of blood. For a few days he experienced some difficulty in swallowing, but under the care of the parish surgeon this passed off, and he apparently recovered. The little patient now continued comparatively well until a fortnight prior to admission, when the dysphagia returned, and gradually increased in severity. The history continues as follows:

He now complains of pain in the throat and epigastric region. He swallowed a quantity of beef tea two days before admission, but has taken nothing since, though he does his best to get something down, and overcome an obstruction which evidently exists. There is nothing to be seen on looking into the throat. The bowels are exceedingly costive, as might be expected.

February 4th.—The child being extremely feeble, and evidently very seriously ill, Mr. Cock saw him with Dr. Addison, with the view of resorting to some operative measure to relieve his present urgent symptoms; but it was not thought advisable to adopt any at this time. Injections of beef tea and wine were ordered every four hours; and any fluid he chose to ask for, to be given by the mouth.

5th.—He has swallowed the greater part of a cup of milk. Fomentations were ordered to be applied to the throat, and half a grain of calomel was given twice a day.

The report states that on the 15th, he was improving gradually. On the 18th the calomel was omitted.

March 9th.—He seems to have remained apparently stationary as regards his health to this time; but the difficulty of deglutition has evidently increased lately.

12th.—He has not swallowed anything for two days; is very exhausted, and appears fast sinking.

13th.—I saw this patient several times during the progress of the case, and was anxious to perform the only operation that appeared to offer any prospect of relief; but once or twice he seemed to swallow a quantity of fluid with apparent ease, which induced Dr. Addison to think it better to postpone any surgical interference until the present time. Accordingly, on this day the little fellow was placed on a table and chloroform administered. I then made an incision about two inches in length along the outer edge of the rectus muscle, in the left hypochondriac region, commencing at the cartilages, and opposite the space between the seventh and eighth ribs. The muscles and fasciæ were cautiously cut through, and several vessels which bled rather freely, tied; the peritoneum was then exposed, and carefully divided upon a director; coils of small intestine immediately appeared in the wound, but were held on one side, whilst two fingers were passed up to the diaphragm, to find the œsophageal end of the stomach. This part of the operation was attended with some difficulty. When, however, the stomach was reached, it was easily recognised by its thickened appearance and velvety feel; and also by the vessels passing along its greater curvature, and its connexion with the descending portion of the great omentum. An opening was immediately made into it, but a large vessel which was divided required ligatures at the two ends, as they bled profusely; the edges were then stitched carefully to the abdominal parietes by an uninterrupted suture. The rest of the wound in the abdomen was closed by similar means. The operation was now completed. ℥iss. of chloroform was all that was used to produce perfect anæsthesia, and it did not give rise to sickness or any untoward occurrence afterwards. Prior to the operation the little patient lay apparently sleeping quietly; when addressed, he understood what was said, but was not capable of answering; the pulse was 82 and feeble; the breath cool, arms and hands cold, but the legs and body tolerably warm. Half an hour after the operation the pulse was 120, and the boy certainly not more feeble than before.

Half an ounce of milk and egg, and of egg and wine were given alternately every quarter of an hour for the first two hours through a tube passed into the stomach by the wound; after that time the intervals of feeding were longer—every hour—and milk only was given. Nutritive injections were likewise administered every four hours, and retained.

On the day following the operation, the little patient appeared comfortable, and answered questions, but of course was very feeble, with a pulse of 120. The milk diet was continued, and the nutritive injections of beef tea had an ounce of wine added to them; the feeding through the tubes was continued every hour, night and day, unless he was soundly sleeping, which occurred at intervals.

On the Tuesday (the operation having been performed at eleven o'clock on Sunday morning) he seemed very comfortable, without any tenderness of the abdomen, and his general appearance indicated that he was in a most favourable condition, considering the important operation which he had undergone; there was no peritonitis. The wound looked pretty well, though somewhat lacking power; the food evidently disappeared from the stomach, and passed into the intestines. He even attempted to swallow some milk by the mouth, and partially succeeded.

On the Wednesday morning, early, he was very comfortable, and was being fed every hour and a half through the tube, which he seemed to enjoy; nay, he even asked for his "poultice," as he called it, when the time arrived for the nourishment being administered. About ten o'clock a.m., after having been fed, he suddenly complained of great pain over the abdomen; he became collapsed and cold, the eyes were sunken, the pulse almost imperceptible, and quickly passing into a comatose state, he died at two p.m.

The post-mortem examination was made the next day by Dr. Wilks, and is as follows:—

"The body presented an extreme degree of emaciation.

"A large part of the œsophagus was affected by the poison, but more especially its middle, which was much constricted, although a narrow passage still existed through it. The constriction was due to a great thickening and induration of the submucous tissue. About half of the tube corresponding to the middle portion had its walls thus thickened and calibre narrowed; the remainder, including the part above and part below, was differently affected. The latter was healthy, as regards its mucous membrane, walls, and size of tube; the former, that is, the part from the commencement of the œsophagus to the constricted part, was considerably dilated. This dilated part was healthy as regards its mucous membrane, but there was a distinct fold of that structure, forming a kind of valve at the commencement of the œsophagus, and again at its junction with the strictured portion. As regards this latter part, where the narrowing of the tube commenced, the mucous membrane was gone, and there was a small sloughy spot; but below this, although the interior was of a dark colour, no ulceration was present, although the mucous membrane had probably been partly destroyed or abraded.

"The external wound in the abdomen had commenced to slough, and the stitches had loosened from their hold. On opening the abdomen a general recent peritonitis was seen, evidently due to a giving way of the sutures, for some of the contents of the stomach were found within it, and lying between the liver and diaphragm; tender flakes of lymph covered various parts, and there was a little fluid serum. On lifting up the parietes, the stomach became still further separated from them, owing to the readiness with which the stitches came out. The part which had been opened was the middle of the organ, or at the lower part of the anterior surface, and scarcely any adhesion had as yet formed around the incision; the colon lay immediately below, and was adherent to the under surface of the wound, as also was a portion of omentum which had curled up into this position. The stomach, on being opened, contained food, and the mucous membrane was healthy, except being of a dark colour around the opening. The small intestines contained food, and the large intestine fæcal matter."

Should another case similar to the foregoing come before me, I should have no hesitation in adopting a similar treatment. It will be seen that the fatal issue in this case seems to have arisen from the accidental giving way of the sutures during the administration of food on the fourth day, and I think it would be advisable to leave a silver tube constantly in the stomach, until firm adhesion had formed between that viscus and the parietes of the abdomen.

The following woodcut is taken from a drawing made when the body was on the post-mortem table;

it shows the position of the wound, and the extent of the emaciation.

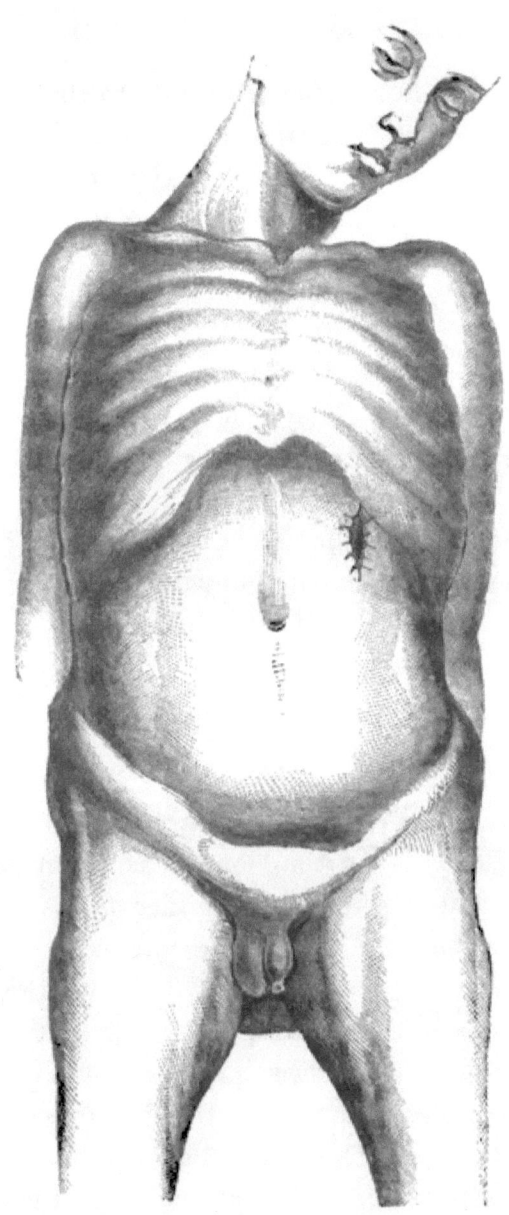

CHAPTER VI.

DISEASES OF THE RECTUM.

LEAVING the intermediate tract to the physician, I come now to the lower extremity of the alimentary canal—the rectum. Here there are four chief diseases which demand our attention: these are piles, prolapsus, polypus, and fistula.

I. HÆMORRHOIDS OR PILES.

These, as seen in the adult, are very seldom met with in children. They do, however, occur. A child may be brought to us with a fringe of vascular swellings, about the size of small beans, all round, within the anus. There has been pain on passing the motions for some few days, with slight oozing of blood. Most probably constipation has preceded. Mild aperients soon give relief, and so far as my experience extends, the piles do not return, or require operative treatment.

II. PROLAPSUS ANI.

This disease is as common in children as piles are rare, and might, in its milder forms, be very easily mistaken for the latter affection. It is, I believe, only the mucous membrane of the bowel that descends in these cases, but the extent to which the prolapsus may proceed is very great; I have seen what might be taken for the entire mass of the rectum, external to the anus. Between this condition and the slightest perceptible protrusion of the mucous membrane, every intermediate degree is met with. The first symptoms of the appearance of this troublesome complaint occur after the child has had a motion. The mother observes that the child is sitting longer than usual, and on inquiring as to the cause, it expresses itself unrelieved. Upon examination, a small portion of the bowel is seen protruding, which, in the milder forms of the disease, soon disappears of itself, or is easily returned. This condition may continue for many weeks, or months; it is generally accompanied by more or less irritability of the rectum, and straining, and sometimes also by irritability of the bladder. Should the child's health improve, or the causes which produce this disease be removed, the affection entirely disappears.

In another and severer form of the disease, a larger portion of the mucous membrane of the bowel descends after each motion, to the extent perhaps of two or three inches, and requires the careful attendance of the nurse or the surgeon to replace it. When replaced, however, it is retained in its position until the next act of defæcation. In this form of the disease severe symptoms of strangulation may occur if the bowel be not replaced. I have seen vomiting, great prostration, and other alarming symptoms in children of eight or nine years of age, suffering from this form of prolapsus. I was once called to the Yorkshire School to see a boy in a cold, clammy perspiration, vomiting, and almost pulseless, who had prolapsus ani to the extent of four or five inches. The prolapsed part was almost livid; it had been down three or four hours, the boy having been afraid to mention the subject. The symptoms entirely disappeared on replacing the protrusion.

In the third form of the disease the prolapsus is constant. The bowel, when returned, is with the greatest difficulty retained, and then only by mechanical means. This form of disease occurs but in the extremes of life, in young children under five, and in old men. The former are always very delicate. The anus is large and patent, and in some cases there seems to be but little power to retain the fæces. The protruded part is generally, although red, paler than natural, and smooth, the rugæ being deficient. The accompanying woodcut represents an ordinary case of prolapsus, not of the severest kind, in a female child.

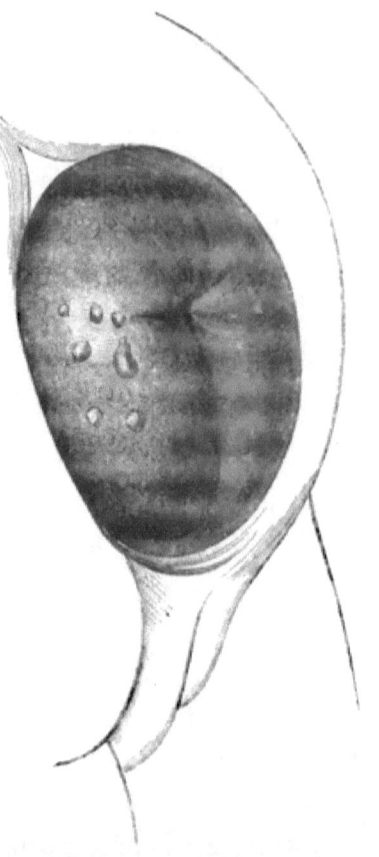

The causes of this affection are, for the most part, the presence of irritating matters in the intestinal canal, worms, and the presence of stone in the bladder. Indeed, such a constant symptom is prolapsus ani of irritation of the urinary organs, or of calculus, that I am often induced to sound the patient if the prolapsus prove obstinate. It occurs also frequently in debilitated children, and in them may be brought on by coughing, crying, or any violent straining, particularly if the bowels are confined.

The *Treatment* of the slighter forms of this complaint, consists in the use of mild aperients; a little rhubarb, soda, and calumba; and copious cold ablutions of soap and water. If possible, it is desirable to have the child's bowels open the last thing at night. And under no circumstances should it be allowed to remain long on the stool. Permanent relief will be given by these means.

In the second form of the disease, we may commence with the above-mentioned measures; but in the course of a few days, tonics, such as iron, or quinine, will be desirable. Astringent lotions (alum, zinc, &c.) should be applied to the protruded membrane, before it is returned. In the third stage, I am in the habit, occasionally, if all these measures fail, of giving the sixtieth of a grain of strychnia to a child of four years old, two or three times a day; and smearing the protrusion daily with the solid nitrate of silver, a proceeding which is unattended with pain. I recommend, also, that the child should be removed into better air, if practicable: that of the seaside, where it can likewise have the advantage of cold bathing, is the best. A pad, with a small hole in the centre, should be applied to the anus; this will allow the passage of the fæces, and at the same time support the sides of the orifice. Great ingenuity and care are required upon the part of the attendant, nurse, or mother, to keep this apparatus firmly fixed, which should be done by straps of "elastic" over the shoulders. Should these means not succeed, it may be advisable to snip out three or four pieces of loose skin and mucous membrane about the anus, radiating from the orifice outwards; under chloroform, of course.

To this proceeding, however, in children, I am averse, as the consequent contraction, when the child grows older, may lead to very unpleasant effects.

III. POLYPUS OF THE RECTUM.

This is a disease in children which, though well known to those who are much in the habit of seeing these little patients, is hardly recognised by many others. It is usually confounded with piles. My attention was first called to it about twelve years ago, by my friend Dr. Willshire, who sent me a case of troublesome bleeding from the rectum in a child. Previous to that time, I confess, I had not observed it. Since then, however, several cases have come under my observation. The mother usually says that the child is constantly passing a little blood, and when it goes to stool, a smooth red body sometimes appears at the orifice, and will occasionally protrude. I have sometimes been told that the child has its bowel down, but the case is distinguished from prolapsus by the bleeding which constantly accompanies it. The straining, and irritability of the rectum, are like that in prolapsus, arising as they do from the same cause, the presence of a swelling in the gut.

These polypi are of a fibro-cellular structure, and exhibit under the microscope its characteristic appearances. They are to be discovered by exploration with the finger, and when there is bleeding from the bowel, this should always be had recourse to. They are attached to the mucous membrane, about an inch or two above the anus, by a pedicle, which can easily be felt by the finger. The treatment consists in tearing them off by the nail; or if this be not possible, in tying them: very slight bleeding occurs. We have not the means of judging whether they grow again from the same spot, but it sometimes happens that two or three polypi form in succession.

For the most part, these cases are brought to the surgeon early, before the loss of blood has become serious. But it once occurred to me at the Infirmary, that a child about three years old was brought in, pale, exsanguine, and almost lifeless. The mother did not know, in this case, of any loss of blood. I felt convinced, however, that bleeding was taking place somewhere; and upon examining the rectum, a polypus of soft, almost fungoid, character, was found attached to the anterior part of the bowel. It was as large as a nut. The mother stated that the child always used the common privy, and so the bleeding had escaped observation. I tied the polypus, and the child became perfectly well.

IV. FISTULA IN ANO.

This disease is much more frequent in children than is commonly imagined, and may occur at a very early period. I have met with it in an infant of fifteen months old. As in the adult, there first appears, in one or other ischio-rectal fossa, a small hard lump, which, as it increases, gives rise to a great deal of pain, and sometimes to much constitutional disturbance. It suppurates very rapidly; the surrounding parts soon become much swollen, the skin becomes red, and quickly gives way. The contents of the abscess are sometimes highly offensive. During its formation the child complains of great pain in passing its motions, and these are sometimes slightly tinged with blood. Great relief follows the discharge of the abscess, and the swelling subsides; but there remains a small opening within an inch of the verge of the anus, and upon pressure, a hardened track may be felt leading towards the bowel. This is the fistula. The opening in the skin is sometimes very prominent, and there appears a small projecting granulation, such as we see when a foreign body is situated at the bottom of a wound. Sometimes there is a small pinhole opening, scarcely perceptible, and requiring great care to discover it. Occasionally a scab forms over the opening, and in this case it gives way once or twice in the twenty-four hours, and permits the egress of matter.

These children are usually said by the mother to have a boil, and are often treated accordingly. The following case shows the uselessness of such a proceeding:—C. S., aged twenty-seven months, was brought to me on the 31st of October, 1851. The mother stated that when the child was four months old, there appeared, as she thought, a boil on the right side of the anus. This quickly discharged, and then closed for about two days, but ever since has continued open at intervals. Poultices, ointments, lotions, &c., had been applied during the whole twenty-three months that the disease had lasted. Upon examination with a probe, this "boil" proved to be a fistula communicating with the rectum. I at once divided the sphincter along the track of the fistula, and in two months' time the child was quite well.

Sometimes, as in adults, there is a communication with the rectum without any external opening. Upon pressing the hard track felt in these cases, we can often cause matter to ooze out by the anus. The following was a case of this kind:—A. B., aged fifteen months, was brought to me in May, 1852, for a swelling on the right side of the anus. It was soft and elastic, but the skin had not quite given way. Upon pressing the swelling, matter appeared at the anus. I at once divided the sphincter and abscess, from within outwards. The child rapidly recovered.

The *Prognosis* in these cases is most favourable, if they are properly treated; but care must be taken to divide the external sphincter entirely, or the surgeon may be mortified by finding the fistula still existing after the wound has healed. There is no reason to apprehend in the case of children any disease of the internal organs.

The *Treatment* of this affection in children is the same as in the adult. The patient having been placed under the influence of chloroform, a flexible grooved probe (shown in the accompanying woodcut) should be passed into and along the fistula, and through its intestinal opening into the rectum. The forefinger of the other hand should next be introduced into the bowel, and the end of the probe bent downwards, and brought out at the anus. The bistoury should then be passed along the groove, dividing everything until the probe falls out. In the after treatment, a piece of lint should be kept in the wound for the first twenty-four hours; after that, cleanliness alone is necessary. The cure is completed in six weeks or two months.

In the following case, in consequence of the sphincter not being divided sufficiently, the operation had to be repeated. A girl, aged two years, was brought to the hospital in May, 1857, with an abscess on the right side of the anus, communicating with the rectum. It had burst a day or two previously. My dresser attempted to divide the sphincter, but failed to do so completely. At the end of three weeks the incision had healed, but a week afterwards the fistula still existed. I placed the patient under the influence of chloroform, and entirely divided the sphincter. At the end of six weeks the child was quite well.

If we are fortunate enough to see the patient before the abscess has burst, and it is close by the side of the anus, although it has not penetrated the rectum, we may be perfectly certain that a fistula will form,

H

and that the first opening externally will not avert it. The best plan, therefore, is at once to divide the abscess and sphincter together. And this is best done by means of what is called Brodie's knife (shown in the accompanying woodcut), passing the finger up the rectum with the knife upon it, and cutting freely outwards into the ischio-rectal fossa, which may be done with impunity, owing to the large amount of fat and cellular tissue therein contained.

CHAPTER VII.

AFFECTIONS OF THE TRUNK.

I. DISEASES OF THE NECK.

THERE is hardly any affection more frequently met with in children than *Enlargement of the Glands of the Neck.* This is in many cases evidently due to some source of irritation about the head, as diseases of the scalp, eruptions on the face, or inflammation of the meatus of the ear. Or it may arise from irritation on the upper and fore part of the chest, some of the lymphatics of that region passing to the glands of the neck; or, though this is rare in children, from inflammation of the throat within. This class of cases differs materially from the forms of enlargement which arise without any ascertainable external cause, and are evidently owing to some constitutional defect. Swelling of the glands from irritation usually disappears when the cause is removed, and does not demand any further treatment. The enlargement that arises from constitutional causes, on the other hand, is one of the most troublesome and obstinate affections with which we meet in early life. Neither rich nor poor are exempt from it, though the most aggravated cases occur amongst some of the badly fed London children. Indeed, wherever children are confined to large towns, this disease prevails. It continually follows scarlet fever or measles. It occurs, also, in what are called strumous or scrofulous children, but these terms convey very little real information; and it is by no means a general rule that the enlargement of these glands depends in any degree on true tubercular deposit. It is a question whether enlarged glands are not more frequent in the children of parents who have suffered from syphilis than in others. It is my own impression that they are.

There is some interest in the *diagnosis* of this affection, since there are other diseases with which it might be confounded. Congenital cysts, filled with fluid, are sometimes met with in the same parts of the neck, and being deeply seated and bound down by the cervical fascia, give a very similar impression to the touch; but by careful manipulation the presence of the fluid within the cyst may generally be detected; the history, also, is different, and the diagnosis may be made certain by an exploring needle. The true type of nævus, viz. the subcutaneous form, also sometimes resembles an enlarged gland, but it has not the same firm hard feeling. It is, besides, usually irregular in shape, whereas the gland is defined and oval. Moreover, the contents of the nævus may generally be squeezed out, and there are frequently enlarged veins over its surface.

Should an enlarged gland inflame and suppurate (which will often happen), it may be at first mistaken for diffused phlegmonous inflammation of the neck; but in the course of a few days the inflammation concentrates itself in or around the gland, and its true character becomes evident. Sometimes, also, small abscesses form in the cellular tissue of the neck, quite unconnected with the glands. These are generally acute in their progress.

The extent to which enlargement of the glands may proceed is sometimes enormous. The subjoined

woodcut represents one of the worst cases I have met with. The patient was a girl aged five years, admitted under my care at the Infirmary for Children. The disease commenced in a mass of enlarged glands on the left side of the neck; these gradually increased, and spread, until both sides of the neck became enormously swollen, and towards the last the swelling involved also the upper part of the chest. The child died apparently suffocated. The circumstances of the patient were tolerably comfortable; she was of ruddy complexion, and appeared to be in perfect health throughout the whole period during which she was under my observation, which extended over two years and a half. Tracheotomy would have suggested itself as an appropriate operation in this case, but the immense swelling rendered it physically impossible. I could not obtain a post-mortem examination.

The *Treatment* may be divided into local and constitutional. In respect to the local treatment, we cannot too constantly keep in mind that these glands become enlarged, very often, in children possessed of many personal attractions, and by whom any permanent scar or other mark of surgical treatment would be felt in after life as a serious evil.

Leeches, therefore, should never be applied, nor should blisters, nor even strong tincture of iodine. Of course, this is of less consequence in a boy; but I am of opinion that all such local remedies are of very little avail.

The constitutional treatment, and in this should be included the general hygienic management of the child, is of the chief importance. As far as medicine is concerned, cod-liver oil and steel wine carry off the palm. The one is a nutriment, and the other a stimulant. Syrup of the iodide of iron is sometimes of use.

The hygienic treatment consists in thoroughly washing the children with soap and water every night, clothing them in flannel from neck to ankle, good diet, and plenty of out-door exercise, be the weather warm or cold. The flannel clothing should, of course, be thin in summer, but I recommend its use in all seasons. If it be possible to send the child to the sea-side in the autumn, it would be well. Of course, country air is at all times desirable.

Should suppuration occur, the best plan of treatment is to open the abscess at the latest possible period, immediately before it would burst. By this means, the desirable result of having the scar as small as possible is attained. The incision should be merely the breadth of the lancet, and always in the direction of, and not across, the fibres of the platysma myoides. The abscess should not be pressed. A piece of wet lint, covered with thin gutta percha, should be afterwards applied. Sometimes these abscesses press upon the trachea, in which case they should be opened at once.

II. CYSTS IN THE NECK.

Cysts of a curious kind become developed in the neck before birth, which I cannot pass over without a few words. Although they are not frequent, the fact of their existence should be borne in mind by the surgeon. They are sometimes attributed by the mother to fright or fancy on her part, as in the following case:—

M. J. B., aged one month, was sent to me by my friend Dr. Willshire, in September, 1852. The mother supposed that at the fourth month of pregnancy she was frightened. Nine days after the birth of her child, she discovered a firm, hard tumour, as large as a walnut, over the sterno-mastoid muscle of the right side, at its attachment to the clavicle. It was very slightly moveable. A seton was passed through it, and kept in for three or four days. Suppuration took place, and the child was quite well in a fortnight.

These cysts are sometimes disregarded by the parents at first, and are not brought under the surgeon's notice until the child is some years old. They may then be taken for chronic abscesses, &c., unless a very careful history has been obtained.

They generally remain stationary for a long time; but if they do increase, they grow very rapidly. They do not cause pain or inconvenience, except by pressure on the neighbouring parts. Their contents are a clear viscid fluid, and the cysts may be either single or multilocular. When superficial, they may be dissected out; but by far the best treatment is to pass a seton through them, and let them suppurate.

III. TORTICOLLIS, OR WRY-NECK.

True wry-neck is apparently a spasmodic affection of the sterno-mastoid muscle of the one side, or paralysis of the sterno-mastoid of the other. The face is drawn towards one shoulder, and forcibly maintained in that position by the muscles. Any attempt to straighten the neck is attended with distress to the patient. When a case of this kind is presented to the surgeon's notice, it should be carefully diagnosed from disease of the cervical vertebræ, the symptoms of which affection will be mentioned hereafter. Wry-neck, if supervening suddenly, is frequently owing to irritating matter in the intestinal canal, which a dose or two of a mild aperient will quickly remove. At other times, it appears to be owing to general want of power, and I have cured it by the use of the sesquioxide of iron. It has been recommended that the muscles should be divided. I have never met with a case in which such a proceeding would have been justifiable. Frictions of the neck are useful.

IV. DISEASES OF THE UMBILICUS.

One can easily understand why this region of the body, which has recently played such an important part in the animal economy, should be liable to various diseases in early life. Of these, the chief are ulceration, either of a specific or simple character, vascular growths, and fæcal fistula.

The specific ulceration will be treated of under the head of syphilis. The *simple ulceration, or fissure*, forming about the umbilicus, requires scrupulous attention to cleanliness, and a slight astringent lotion. It only occurs in children that are neglected, unless psoriasis is present in other regions of the body, in which case it seems to be a part of the more general affection.

V. GROWTH FROM THE UMBILICUS.

I have seen, in three or four instances, a small polypoid growth from the umbilicus in children from one to eighteen months old. This growth appears at first like a granulation of unusual size, but I have seen it attain a length of three-quarters of an inch. In these cases it constitutes a pendulous excrescence, which might almost be taken for a portion of the cord. Sometimes it appears sunk within the umbilicus, and requires some dexterity to expose its base. It is of red colour, and pours forth a constant purulent secretion; occasionally it bleeds, but gives rise to no pain or other inconvenience. I have not been able satisfactorily to trace the history of these cases, but I presume they result from imperfect separation of the umbilical cord. The mother generally states that the growth appears after the cord comes away. It is shown in the accompanying lithograph.

The first case that I saw was a child seven weeks old, sent to me by my friend Mr. Marshall, of Mitcham, in November, 1852. He was a strong little fellow, but growing from the umbilicus was a pendulous oblong body, which had been there since the separation of the cord. My note-book says, "appearing like a long hæmorrhoid." The mother stated that when the child cried, the tumour became of a much deeper red colour than when it was quiet.

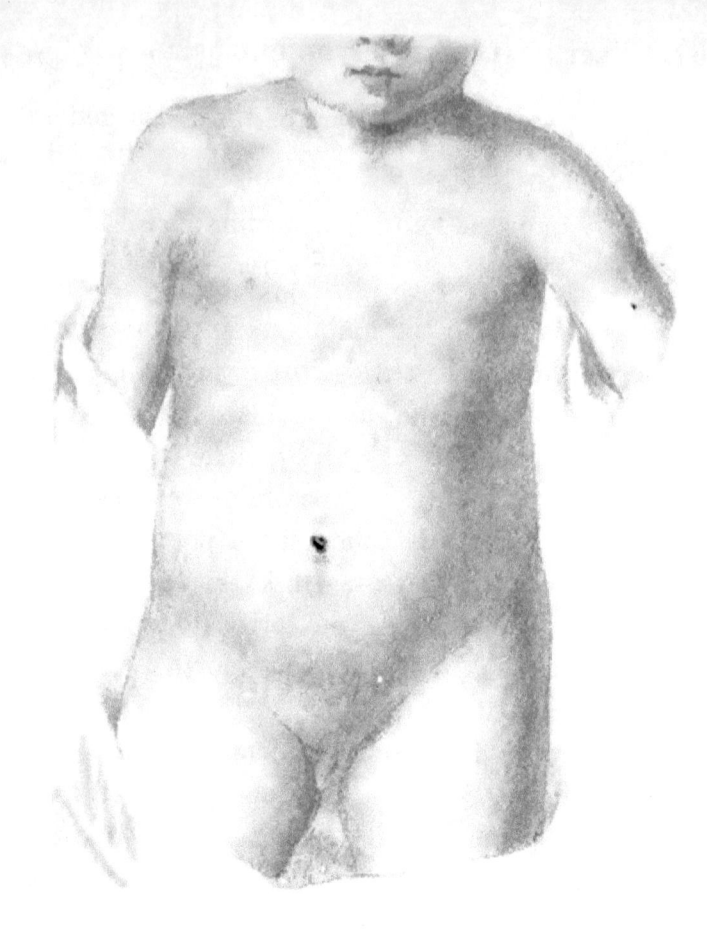

It was constantly pouring out a purulent secretion, which kept up a continual irritation. Dry lint, nitrate of silver, and astringent lotions, had been applied to it without avail. I encircled its base with a ligature, which came away in twenty-four hours. The child did not suffer any pain or constitutional disturbance, and at the end of a week it was quite well. Under the microscope, the growth appeared to be composed of fat cells, and to contain no vessels.

Another case occurred in a child a year and a half old; the excrescence had existed from one month after birth, and had been treated by various practitioners with sulphate of copper, nitrate of silver, bandaging, &c. It had grown to the extent of nearly an inch in length, and the abdomen was excoriated by the discharge. A ligature was firmly applied around it, and in a week the child was quite well.

The entire treatment consists in the application of the ligature. The growth separates in twenty-four hours, and the wound heals rapidly.

VI. FÆCAL FISTULA AT THE UMBILICUS.

The two following cases of this rare affection have occurred to me, and it will be seen that both were benefited by the treatment adopted. The first case came under my notice while I was reporter to the late Mr. Aston Key:—J. C., aged fourteen months, was admitted into Guy's Hospital on the 5th October, 1842. *History.*—His mother stated that ever since he was born there had been a discharge from the umbilicus of a thin yellow colour, and of a faint odour. The umbilical cord did not come away until the eleventh day. There was then a glutinous discharge, which fixed the parts together. A poultice was constantly applied for three months, and the nitrate of silver occasionally, but the discharge still continued. The edges of the wound had been pared, and pieces of strapping had been applied to bring the parts into apposition.

The child was fat and healthy, with fair complexion. The aperture of the umbilicus was open, and communicated with a sinus, into which a probe could be inserted to the depth of two inches. The surrounding parts were excoriated.

Three days after admission, chloride of zinc, made into a paste, was introduced, on the end of a probe, into the sinus The discharge disappeared for two or three days, but then returned, and on the eighteenth day after the application, there was again a copious flow, of a yellow colour. Four days afterwards the discharge was described as resembling fæces. Accordingly, on the 29th of October, the edges of the opening were pared, and the freshly incised parts brought together by means of three needles and straps of plaster. After three days the needles were removed, and plaster and a bandage applied around the abdomen. On the 12th of November the child was discharged, with the wound almost healed.

The second case occurred at the Royal Infirmary for Children, under the care of Dr. Willshire and myself. A little boy, aged five, was admitted in December, 1857, with a fistulous opening at the umbilicus, large enough to admit the little finger. This had existed three weeks, and was believed to have been caused by tubercular inflammation (?) connecting the colon and abdominal wall, and finally suppurating externally. The child was in a miserable plight, the fæces escaping to the amount of several fluid ounces daily, and very small quantities passing by the anus. The fæces from the umbilicus had an offensive odour, and were of such a character that little doubt could be entertained that the bowel involved was the transverse colon. The edges of the fistula were pared, and the adjacent skin slightly dissected up, to allow the closure of the aperture. The parts were accurately retained in position by a quilled suture, with three ligatures. On the second day after the operation a good motion passed *per anum*, and until the fourth day no fæces whatever had escaped by the wound. The sutures were then removed, and it was found that union was complete, except in a track scarcely large enough to admit a probe, and through which a few drops of almost colourless fluid escaped daily. All the fæces continued to pass *per anum*, and the child left the hospital much improved in health.

The only efficient treatment of this affection is that which was adopted in the above cases. And this is not always successful. In the case last reported, though the child went out pretty well, I have since heard that the symptoms have again increased. Nor, when we remember the anatomical condition of the parts involved, how near to the abdominal parietes the posterior wall of the bowel extends, and how small and crooked a passage is left for the fæces when the external opening is closed, can we be surprised that this should be the case. Any one who should devise a means by which a more perfect restoration of the natural canal could be effected, would confer, upon all sufferers from fæcal fistula, a great boon.

Paracentesis Abdominis.—I cannot leave the subject of the abdomen, without first referring to the fact that one may occasionally be called upon to perform paracentesis at a very early age; though of course only with a view to temporary relief. I was requested by Dr. Hutton, at the Infirmary for Children, to see a child, aged one year and nine months, with ascites, attributed to tubercular peritonitis. The child was much emaciated, and the abdomen was enormously distended. There was no disease of the lungs, or liver, or heart, and the urine was free from albumen. Three months previously the child had had scarlet fever. About a month afterwards the abdomen began to enlarge, and had continued to do so ever since. The treatment pursued prior to admission was purging with salines, and the administration of diaphoretics. Latterly calomel and Dover's powder had been given, and mercurial ointment rubbed into the abdomen. She was admitted into the Infirmary on the 22nd Nov., 1859, and then measured round the abdomen twenty-eight inches. The mercurial treatment was continued without relief, and on the eleventh day from admission, the child's distress being extreme, the fluid was drawn off from the abdomen. The operation gave great relief, but in two days the measurement was again twenty-four inches, and in two days more it had increased to twenty-six, when the child was removed from the Infirmary by her mother. In a few days she died. During the whole time she was in the Infirmary her appetite was most voracious.

The operation of tapping in a child is the same as in the adult. It is of course necessary, however, to use a smaller trocar and canula.

VII. DISEASE OF THE VERTEBRÆ—ANGULAR CURVATURE.

This disease may occur in either of the three regions of the body: the cervical, dorsal, or lumbar. The pathology of the affection is, of course, always the same: but its symptoms and effects vary in the different regions. The very earliest indications of this disease are seldom brought under the surgeon's notice; for the most part when a patient is brought to us, the symptoms are already somewhat advanced. This is more especially the case when the disease is seated in the lumbar region. When the *cervical* vertebræ are affected, there is first a peculiar stiffness and uneasiness about the whole region of the neck, and an indisposition to turn the head to either side. The muscles seem to be on the alert to restrain any motion, almost as if aware of the danger of a sudden movement. There are also obscure pains about the neck, extending over the occiput if the upper vertebræ are affected; or downwards over the shoulders and along the upper extremities, when the lower vertebræ are involved, which is the least frequent form of the disease. On examination, there are found thickening and tenderness in one particular spot at the back of the neck, generally high up over the third or fourth cervical vertebra. There is also a great fulness about the junction of the neck with the occipital bone, which is very characteristic. It is due to œdematous effusion in, and hypertrophy of, the cellular tissue around the deep-seated muscles.

At this stage of the affection, the disease has been taken for a case of torticollis merely. But in wryneck the head is more drawn to the side, and there is an absence of the local tenderness so characteristic of diseased spine. Moreover, when the child is lying down and wishes to rise, if the vertebræ are diseased, it invariably puts its hand to its head, and lifts it up. And if the surgeon places his hand gently on the top of the child's head, he finds that it cannot be ever so slightly rotated without intense pain.

As the disease advances the patient becomes unable to rise from the bed; having more or less paralysis of the extremities. An abscess forms in some part of the neck, and so far as my experience goes, invariably points in the pharynx. Occasionally the disease implicates the articulation between the first and second cervical vertebræ, and then a sudden movement of the head may cause instant death.

Deformity is exceptional in disease of the cervical vertebræ. If it occur, the convexity of the curvature is more frequently forwards than backwards; so that the head appears somewhat thrown back.

When the disease occurs in the dorsal region, on the other hand, very frequently the first symptom noticed is a projection of the back. Upon inquiry, however, the child is found to have been out of health for some weeks, or even months; languid and indisposed to run about—always a suspicious circumstance in children. There is also, in the early stages, more or less pain about the front of the chest and upper part of the abdomen; that is to say, at the terminal extremities of the dorsal nerves. And too much stress cannot be laid upon the rule that, whenever a child suffers from vague pains affecting the front of the chest on both sides of the mesian line, it should always be stripped, and the spine examined. Upon minute investigation, it will be found that these pains, when arising from diseased vertebræ, are more or less symmetrical on the two sides.

At a very early period, there may be local tenderness on pressure, but the absence of this symptom is by no means a sign that there is no vertebral disease. Concussion, as from jumping, may occasion signs of pain, but this is not much to be relied upon.

As time goes on, unless the disease should be arrested, the angular projection becomes more marked. The ribs, therefore, are necessarily brought into closer contact, and the viscera within the chest suffer; the deformity causing embarrassment of the circulation and respiration. The spinal cord may or may not be affected. In many cases, it seems wonderfully to accommodate itself to the altered course of the canal. In others, it becomes compressed, and paralysis of all the parts below takes place, with the usual train of wretched symptoms connected with loss of power of the bladder.

Even at this stage, the disease may be arrested; as is evidenced by the dwarfs who are so numerous, with the head sunk, as it were, between the shoulders, and with the projection known as humpback. In these cases, the spinal cord seems to become adjusted to its new position, and its functions gradually return.

It is remarkable, that abscess is much less frequent in disease of the dorsal than of the lumbar vertebræ. When it does exist, there is, for the most part, no curvature, and when the deformity is greatest, there has often been no external abscess at all. The matter usually presents at the back; in the worst cases, however, it is apt to run along the rib, to the side of the chest, and may discharge itself into the pleura.

Lastly, the lumbar vertebræ may be affected; in which case, abscess almost always forms. As in the dorsal region, the projection is often the first thing that attracts the mother's attention, but an abscess very generally accompanies it. This abscess may point either in the back, in the sides of the abdomen, or, which is most usual, in the thigh: either the fore and upper part, or more rarely at the back of it.

Psoas Abscess.—When the matter points in the thigh, it is called psoas abscess. This form of abscess is invariably connected with disease of the lumbar vertebræ. As is well known, it runs down the sheath of the psoas muscle, and hence its name. The first symptom is pain in the fore part of the thigh, or the lower part of the abdomen: in fact, in the course of the nerves emanating from the lumbar plexus. We frequently, also, find a deep-seated fulness over the psoas muscle, which may be felt by relaxing the abdominal parietes. These symptoms, however, are not always to be observed in children, they being unable fully to explain their feelings; and the first thing discovered may be swelling in the front or back of the thigh. This swelling, extending as it does under Poupart's ligament, and to a certain extent bisected by it, may be made, by pressure, to pass up and down in the course of the psoas muscle, and may be felt either above or below the position of the ligament. There is usually an impulse on coughing. If the abscess point at the back of the thigh, this last-named symptom becomes of great importance, as it at once distinguishes this affection from abscess connected with hip-joint disease. There is also, at an early period, more or less pain and difficulty in perfectly extending the thigh.

When the abscess is connected with disease of the lower lumbar vertebræ, it most frequently finds its way to the surface beneath the lower border of the glutæus maximus; passing out of the pelvis with the sciatic nerve.

These are the symptoms which characterize disease of the vertebræ, according to its seat. The mischief, wherever seated, commences first of all in the intervertebral substance. By some, it is said to be a tubercular deposit there; which is certainly not the case. The intervertebral substance diminishes; becomes absorbed, apparently; as it may entirely disappear, leaving no trace of any abscess. The bones are secondarily affected. Sometimes it appears as if disease commenced first in the cancellous tissue of one of the vertebræ; but upon closer examination, it will be found that the intervertebral substance has gone, and that two vertebræ have become joined together, as shown by Dr. Wilks in his work on pathology. As the disease advances, the bodies of the vertebræ may disappear, leaving only the arches and spinous processes, which form the peculiar projection in the dorsal region. Any number of the bodies of the vertebræ may be removed in this way, and at last, if the disease be arrested, synostosis occurs between the bodies of the unaffected vertebræ above and below. As many as six or seven vertebræ have been known to be absent; and the third or fourth dorsal vertebra has been found anchylosed to the eleventh. In these cases, paralysis will be present, if the disease has been rapid in its course: if its progress has been slow, so as to allow the spinal cord time for accommodating itself to the change, paralysis may be escaped, even when the projection is the greatest.

The *Prognosis* differs according to the region affected, and also according to the period at which treatment is commenced, and the state of the child's constitution. It is not very unfavourable if the child is seen early, and can be placed in suitable circumstances, except in the case of psoas abscess, which I always regard with great suspicion. The severity of the disease in this case, I believe, is owing to the length of the track through which the matter passes. Next to these cases, the most unfavourable are those occurring in the upper part of the neck, when there is a special risk from sudden motions of the head. Much will depend upon the exactitude and perseverance with which the little patient's friends or nurse carry out the directions hereafter given, and which they are very prone to think needlessly tedious and severe. Even when death eventually results, the length of time through which the disease may be protracted is often extraordinary.

The grand principle of *Treatment* in all cases, is rest in the recumbent posture, in order to relieve the diseased part from the weight of those above. In the cervical region (the child lying quite flat) the head should be firmly fixed by means of a couple of sand-bags placed beside it. The young patient will soon experience the comfort of not being allowed to move it, and will submit willingly to the restraint. The neck should be accurately supported. The length of time during which it is necessary to continue this plan varies considerably; weeks or months are always required. When these patients are allowed to get up, it is necessary that the weight of the head should be taken off the neck by a mechanical contrivance, having fixed points on the hips below, and at the occiput above, and a support around the latter. The admirable instrument contrived by Mr. Heather

118   SURGICAL DISEASES OF CHILDREN.

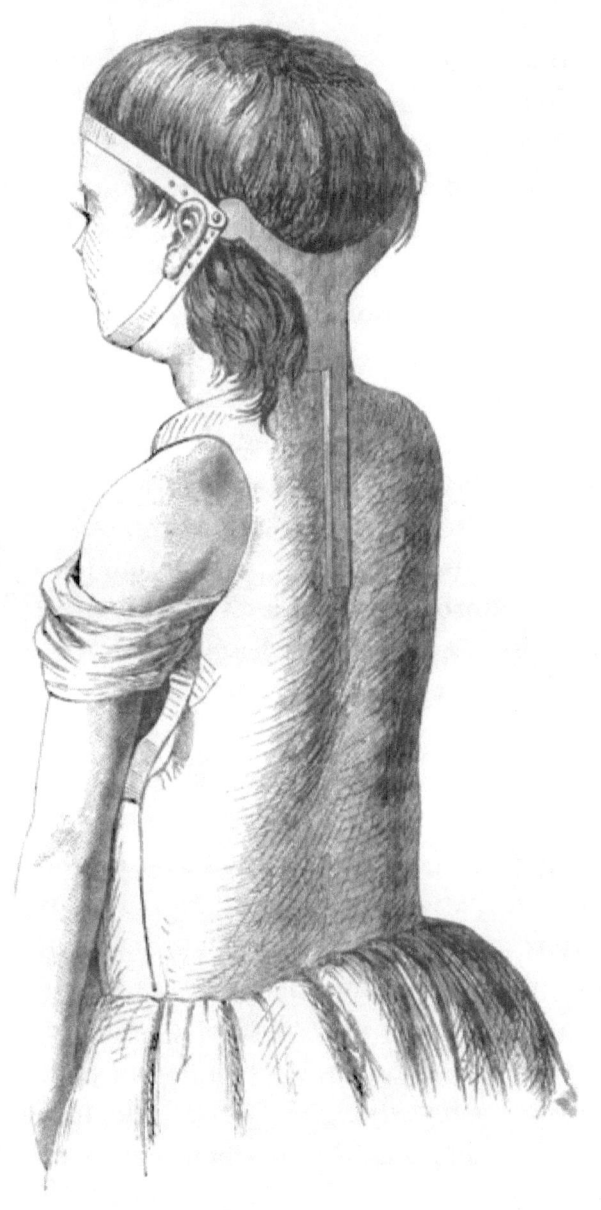

Bigg, is shown in position in the accompanying woodcut.

When the disease affects the dorsal or lumbar region, the child should be allowed to lie in the position most comfortable to itself. The object to be aimed at is to obtain anchylosis, and therefore no attempt should be made to prevent deformity; the approximation of the unaffected vertebræ being a necessary condition of the cure. Mechanical treatment, in my own opinion, is of no avail in this form of the disease. No doubt if a stiff iron bar could be rivetted through the bodies of the vertebræ, it might be very efficient! but inasmuch as there is no fixed point above (the shoulders being moveable), I do not see how it is possible to prevent motion of the spine by an external apparatus. Nay, it seems to me worse than useless, the patient having to support the additional weight of the instrument.

With the view of preventing the formation of abscess, various means, as issues, moxas, or setons have been frequently recommended. I have tried these, and the patients have done well. But I have also omitted them, and the patients have done as well; nay, I believe better, as not being subjected to the irritation of the sore, and the continual disturbance of the dressing. I have therefore quite discarded the use of these means; but I sometimes employ blisters, not with the idea of remedying the disease, but to relieve pain, if it be complained of, for which purpose they seem to be useful.

Can anything be done to prevent the formation of

abscess? Certainly not, so far as regards any means directed to this end apart from the general care of the disease. If abscess forms, the great question arises: should it, or should it not, be opened? For the most part, decidedly it should not. But if it be so large as to cause great inconvenience, or there be much pain connected with it, or the little patient's friends are anxious that "something should be done," it may be opened by a valvular incision. This sometimes gives great relief, and is very seldom attended with any ill results. The skin should be drawn as far upwards as possible by the flat hand, and then a double-edged knife plunged through the skin, thus made tense, at once into the abscess, or a trocar and canula may be used. When the pus ceases to flow, the hand should be removed, and the opening into the abscess will be closed by the skin.

The issue of the case seems to depend very greatly upon the place at which the abscess points, whether near the spine, or at a distance from it. In the former case the results are often favourable; in the latter they are seldom so. In the majority of instances of psoas abscess, the patient, though it may be after the lapse of years, eventually succumbs.

During the whole treatment of all these cases, a generous diet should be given, since nature has much to repair. A pure and dry atmosphere, and a well-ventilated apartment, are very important. The wards of a large hospital, though by no means in themselves desirable, are infinitely superior to the homes of the poor. The sea-side is of all places the best. Some of the many well-contrived beds which are now in use, for moving the little patient without exertion on his part, and for maintaining cleanliness, are to be recommended. The medicinal remedies are those which are useful in all this class of diseases—cod-liver oil, iron, quinine. The latter remedy especially, where there is want of appetite, is useful. It is very undesirable to dose these children with mercurials. The bowels may be confined for two, three, or four days without disadvantage. If, in consequence of the distortion of the spine, any inflammatory complication arise in respect to the organs of the chest, little can be done towards arresting the mischief, beyond placing the patient in the most generally favourable circumstances.

Psoas abscess may open into the bladder or intestinal canal, and in the latter case gives rise to the appearance of diarrhœa; this should be carefully distinguished from true diarrhœa, which is a frequent concomitant of disease of the spine, especially in the later stages. An examination of the stools easily decides this point.

CHAPTER VIII.

URINARY AND GENERATIVE ORGANS.

THE importance of the diseases of these organs in both sexes can scarcely be exaggerated. The happiness of many individuals is so bound up in the perfection of the functions to which they are subservient, that too much consequence cannot be attached to their right management in early life. I shall speak first of these organs in the female, and then of them in the male.

DISEASES OF THE URINARY AND GENERATIVE ORGANS IN THE FEMALE.

In both sexes these organs undergo so considerable a change as the period of puberty approaches, that their anatomy differs somewhat in the child and in the adult; and before proceeding to describe their diseased conditions in children, I shall briefly refer to the structural peculiarities which they present. In the female these are, for surgical purposes, comparatively unimportant. The labia minora are of great relative size compared with the labia majora, the latter being almost hidden by the former; and the urethra has a course more directly upwards, in consequence of the bladder being situated higher up, so as to lie in the abdomen rather than in the pelvis. Towards the age of seven or eight years the parts begin gradually to assume the form which they possess in adult life.

I. DISCHARGES FROM THE VAGINA IN CHILDREN.

Of these discharges there are two kinds—Gonorrhœa and Leucorrhœa.

That such a disease as true *Gonorrhœa*, communicated by the foul contact of some person affected with that disease, does occasionally occur in young girls of four years old and upwards, cannot be denied. But to distinguish it by any pathognomonic symptoms from some cases of infantile leucorrhœa, is, I am bound to maintain, an impossibility. All practitioners are aware that it is the almost universal custom of mothers and nurses, among the poor, to attribute all cases of discharge in young girls to some such cause. Generally they come mysteriously whispering—the child has been dandled on some one's knee, or there is a young man lodging in the house, whose linen is foul. As a rule, not the slightest importance is to be attached to such statements. Infantile leucorrhœa is not a rare disease, and it is highly desirable that the minds of women should be disabused of this kind of prejudice. It is certain that only circumstantial evidence of the most unequivocal kind can lay any basis at all for the imputation. No stress is to be laid upon the symptoms, unless there are evident marks of violence. How is it, one may ask, that in almost all cases in which such assaults are suspected, the disease communicated is gonorrhœa, and not syphilis?

During the whole course of my experience I have met with only one case in which syphilis was communicated to children. The patients were two sisters, aged respectively eleven and twelve, and the young man confessed his crime. This rarity of syphilis appears to me to be a very strong point in the case. In regard to the statements made by the children themselves, very little reliance can be placed on them. They are frequently tutored, more frequently still are cross-questioned, till at last they believe what they have so often heard; perhaps, most frequently of all, the unaccountable spirit of lying, which seems often to infest the infant mind—it may be as the result of dreams—is the entire origin of the tale. And having referred to the dreams of children, I may remark, that a careful observation of their ways will often detect the evidence of impressions upon their minds, which can only be attributed to the confusion of that which has been dreamt with that which has been really experienced. How, indeed, should it be otherwise? How should the child, without the means of judging which we possess, know that that which he has distinctly felt, has never been?

But I need not enlarge on this subject; the admirable articles of Mr. Wilde, of Dublin, in the *Lancet*, and elsewhere, have given sufficient evidence how falsely men may be charged with communicating gonorrhœa to children. At the same time, and for the honour of humanity, we are sorry to be obliged to confess it, there are men, as the case I have mentioned shows, who, under the idea that disease may be got rid of by intercourse with a child, will not scruple to make these unsuspecting creatures the victims of their baseness. Gonorrhœa also may be communicated to children in an indirect way, as the following case shows :—

A married man, aged thirty-two, came under my care with gonorrhœa. In the course of a week, he sent his wife to me; she was suffering from the same disease, but was not cognizant of its nature. Very shortly afterwards, she brought me her three female children, whom she had washed with the same sponge that she herself used, and all three of whom were suffering from a similar discharge. There was a general redness about the vulva, swelling, and a purulent discharge, not very profuse. In one case, there was ardor urinæ. Poppy fomentations were employed for a few days, and afterwards an astringent injection of one grain of sulphate of zinc to an ounce of water. They soon recovered.

The treatment of gonorrhœa in the child is simple. As a rule, poppy fomentations, followed by a mild astringent injection, seem to be all that is required.

Infantile Leucorrhœa is a disease essentially depending upon some cachectic condition of the patient. It may be almost likened to a strumous disease affecting a mucous membrane. It occurs most frequently, but not exclusively, in weak, ill-conditioned children, or in those who are badly fed and badly clothed, and who live in ill-ventilated dwellings. And it is possible that locality may have a share in its production. I have an impression that it is more prevalent on the south than on the north side of London. When it occurs in the well-to-do classes of society, the children are generally what is called strumous. It may occur at any age. The first symptom is a slight mucous discharge from the vagina, which causes redness and excoriation of the vulva and neighbouring parts, sometimes looking like a species of psoriasis. There is pain in passing water. In a few days, the discharge becomes purulent. If first seen at this time, which is most usually the case, the swelling of the vulva has generally subsided, the pain in passing water is diminished; the vagina is red; and there is a constant, profuse, purulent discharge;—do what the mother will, she cannot keep the child clean. The discharge goes on for an indefinite time; often, in spite of the most various plans of treatment, it will last a long while; at other times, it seems easily to yield. The *Prognosis*, therefore, is very uncertain; it depends upon the constitutional condition of the patient. If the child is neglected, and exposed to wet or cold, even sloughing of the vagina may take place.

The *Treatment* that I have usually adopted in these cases has been the administration of the tinct. ferri sesquichloridi, in doses of from two to eight minims, three or four times a day, in simple syrup; using at the same time an injection of the sulphate of zinc or alum, or of the two combined. Sometimes I give six or eight grains of chlorate of potash, in syrup; but there is nothing more valuable than a due attention to cleanliness, good diet, and exercise; which latter, though apparently contra-indicated by the condition of the vagina, improves the child's general health, and should never be neglected.

In tedious cases, it is desirable to vary the injections; the more useful are those of zinc and alum, or glycerine, with liquor plumbi diacetatis, in the proportion of a drachm of the latter to an ounce of glycerine and a pint of water; or chloride of zinc, one grain to an ounce of water; or nitrate of silver, a grain and a half to the ounce. Great care must be taken in the use of the syringe, first to avoid irritating the inflamed vagina, and secondly to insure the thorough application of the lotion to its surface. The syringe most useful is an ordinary squirt (not the female syringe). It is best made of gutta percha, but such an instrument is rather expensive. The hygienic treatment, as in all diseases of children, is, of course, of the utmost importance.

II. STRUMOUS ULCERATION OF THE VAGINA.

I am not prepared to say that this designation is pathologically correct. I apply it to certain cases which I have seen now and then, and which it is of the greatest importance should be correctly diagnosed, as they simulate syphilis. There is found a superficial ulcer on one or other labium, which has extended rapidly till the time of the surgeon's seeing it. This has been, in my experience, generally in three or four days from its first appearance. There is pain in passing water if the urine comes in contact with the sore. The labium is swollen and red. The edges of the sore are studded with bloody points, and there is a great deal of thin purulent discharge. In one patient that I have seen, the sore had acquired the size of a florin. The ulcerated surface heals, however, as rapidly as it spreads, if judiciously treated. The children are always dirty, weak, and ill-conditioned, and the simplest treatment is the best; the grand principle being to avoid any irritating application. With warm water externally and quinine internally, the patients usually soon get well.

It will be seen from the above description how likely this ulceration is to be mistaken for syphilis; indeed, this very mistake has been made under my own observation.

Case.—F. G., a delicate, light-haired child, two years of age, was brought to me in January of this year, with this peculiar form of ulceration on both labia. The mother stated that four days ago she found one labium swollen, and a little raw place on it. This spread most rapidly, and when I saw the child there appeared a very superficial sore, about the size of a shilling, on one labium, and one about half as large on the other. Both labia were much swollen and inflamed, and there was a great deal of muco-purulent discharge. In the course of a week, under cleanliness, water dressing, and vinum ferri, both were well.

III. NOMA.

Under this name there has been described a sloughing of the vulva of a peculiar character. It is undoubtedly analogous to cancrum oris, and, like that disease, seems much less frequent now than formerly. A record of only one case occurs in my note-book.

Case.—L. A., aged nine years, was admitted into Guy's Hospital, under Mr. Cock's care, on the 4th November, 1854, on account of sloughing of the vulva and inability to pass water. This child lived at Plumstead, and there did not appear any reason for the attack. The mother stated that three weeks previously the vulva became red and excoriated, and shortly afterwards began " to look of a curious colour." Upon admission there was a gangrenous sore occupying the right side and upper part of the vulva, pouring out a fœtid discharge. The child was suffering constitutionally; there was constant vomiting, and purpurous spots appeared over the body. Notwithstanding this, there was no emaciation. The treatment adopted was, to draw off the urine three times a day, administering support freely with wine and ammonia. In five days the patient died.

The bladder and lower part of the rectum were inflamed; the gangrene had spread considerably, involving the whole of the vulva. There was diffused apoplexy of the lung.

IV. CLOSURE OF THE ORIFICE OF THE VAGINA.

I can scarcely leave this part of my subject without alluding to the above-named condition. It is congenital. The nurse brings the child, and generally says that it does not pass its water comfortably, and that she cannot properly wash it. The fear is lest the malformation should exist in after life. The mischief seems to be an union, by slight adhesion, of the two sides of the orifice of the vagina; there being a small aperture above, through which the urine, with more or less difficulty, passes. The canal itself is perfect. The treatment is to separate the adherent parts by a blunt probe: very little force is necessary. Of course a few drops of blood are lost, and if care be not taken, the two sides will again grow together. This is to be prevented by touching the surfaces with nitrate of silver, and introducing a piece of lint between them. If the separation be maintained for twenty-four hours, it will be sufficient. I have no doubt that these are the cases which in after life give so much trouble by causing retention of the menses, and are described as cases of imperforate hymen.

V. VILLOUS GROWTH FROM THE BLADDER.

This disease is very unfrequent; only four cases of it, in children under twelve, are on record. These are referred to by Mr. Birkett in his paper in the Medico-Chirurgical Transactions, vol. xli., from which I abbreviate the following notes of a case which came under my own observation, having been admitted into Guy's Hospital under Mr. Birkett's care.

Case.—S. A. J., a girl aged five, was admitted on the 26th of December, 1857. She was a strumous child, and in a state of great exhaustion. Eight weeks before, she had complained of pain in the hypogastric region, and there was a diminution in the quantity of urine passed. After three weeks complete retention had occurred, and the catheter was used. A week later a purulent deposit was noticed in the urine. When admitted, the urine dribbled away involuntarily, but unless the catheter was used the bladder became distended. She appeared as if she had lost much blood. The catheter was introduced as often as required, and iodide of iron was given.

There seemed to be, at first, a slight constitutional improvement, but the difficulty in micturition continued. On the 7th of January, a dark-red growth protruded through the vulva. This interfered with the introduction of the catheter. On the 12th, the child was placed under the influence of chloroform, and the urine drawn off. It was then ascertained that the growth projected through the meatus urinarius, and was attached to the anterior part of the neck of the bladder. The index finger could be passed into the bladder without impediment behind the growth, the projecting portion of which could be partially returned into that organ. A hard mass could be felt behind the symphysis pubis. The new growth was composed of lobes and lobules; it was not very vascular, but like the firmer varieties of nasal polypus. Mr. Birkett passed a strong silk ligature around the root of that portion of the growth which protruded. This ligature separated the parts to which it was applied, but other portions afterwards made their appearance. A few days after this pyrexia supervened, and on the twenty-fifth day from her admission the child died.

On the post-mortem examination there was found pyelitis of the right kidney, but no other internal disease. The urinary bladder was much enlarged. Projecting through the meatus there was a growth about an inch and a half in diameter. Within the bladder were a number of other growths, arising from its anterior wall, and forming together a mass about equal to that which protruded externally. These growths were polypoid in form, and consisted of fibro-cellular tissue.

In connexion with the diagnosis of this case, the only question that arose was whether the child had a stone in the bladder. The use of the sound immediately revealed the soft character of the body occupying that viscus.

VI. CALCULUS IN THE FEMALE.

The general symptoms of stone in the bladder are so much the same in children of both sexes, that I shall defer the description of them until the subject comes under our consideration in the male. The operative *treatment*, of course, is very different.

For the removal of stone in female children, one of two courses may be adopted—either the division of the neck of the bladder and the employment of the forceps, or lithotrity. To the former operation, however, I should only have recourse in those cases in which the latter was inapplicable: for instance, when there existed a very hard oxalate of lime calculus, which resisted the powers of the lithotrite. If compelled to have recourse to division of the urethra, the mode of proceeding will be the same as in the adult. A director being passed into the bladder, a straight bistoury should be cautiously insinuated along it, so as to nick the upper part of the neck of the bladder and urethra. The extent of the eighth of an inch is sufficient. The forceps should then be passed into the urethra, and very gently expanded at the upper part of that canal, to produce dilation of the neck of the bladder. This should not go beyond the least amount that will suffice for the extraction of the stone. The great danger of this operation is, lest permanent incontinence of urine should ensue, which is very likely to be the case if much injury be done to the neck of the bladder. The stone, then, having been seized by the forceps, should be cautiously extracted; or the dilation may be effected by the pressure of the stone, instead of by expanding the forceps; and in this case ample time should be given for the gradual yielding of the tissues.

It has been recommended to employ dilation alone, without incising the neck of the bladder, but this plan is even more apt than the former to give rise to incontinence of urine. By far the best plan of treatment is to employ lithotrity. The shortness of the urethra in girls renders this operation, by comparison, extremely easy. And when we are not dealing with a very hard calculus, the stone will be found readily to give way. In some rare cases, the stone may be of the soft variety known as xanthic oxide, in which case it breaks up like butter under the lithotrite. A case of this kind occurred in Guy's Hospital, under Mr. Birkett's care, about two years ago.

Operation.—The child should lie on a table with her feet resting on either side, her knees well separated, and the pelvis slightly raised; the surgeon sitting opposite to her at the bottom of the table. Chloroform should be administered. The urine having been drawn off, half-a-pint to a pint of lukewarm water should be injected through a gum-elastic catheter into the bladder, until it is moderately distended. The lithotrite, well oiled and warmed, should then be introduced, and the stone seized and crushed. If it have been broken into two or three large fragments the surgeon should again seize these, and break them up, taking care not to do too much the first time. The operation may require to be repeated two or three times, seldom oftener, in young female children. Should a fragment become impacted in the urethra, which rarely happens, owing to the shortness of that canal, it can easily be extracted by a pair of forceps. Serious symptoms scarcely ever arise in these cases; whatever irritation there may be of the bladder soon passes off.

The following case occurred in Guy's Hospital, and was operated on by Mr. Poland:—E. R., aged two years and a-half, was admitted in October, 1847, under the care of Dr. Golding Bird. She had suffered from partial incontinence of urine and pain in micturition for a year and a-half. The symptoms had increased in severity during the last three months, and were accompanied by prolapsus ani and mucous deposit in the urine. The child was put under the influence of chloroform, and Mr. Poland introduced the lithotrite (as shown in the accompanying woodcut). A calculus about the size of a bean was crushed. The debris were extracted by the instrument, which was introduced several times. No untoward symptoms arose; at the end of a week a small fragment was detected, crushed and removed. From this time she perfectly recovered, and obtained complete control over her urine, which she sometimes held for six hours. The patient left the hospital at the end of four weeks; and two months afterwards had no urinary symptoms.

Incontinence of Urine will be best treated of when describing the diseases of the urinary organs in the male; as, although it does occur in girls, it is much more frequent in boys; and the causes and treatment of the affection are the same in both cases.

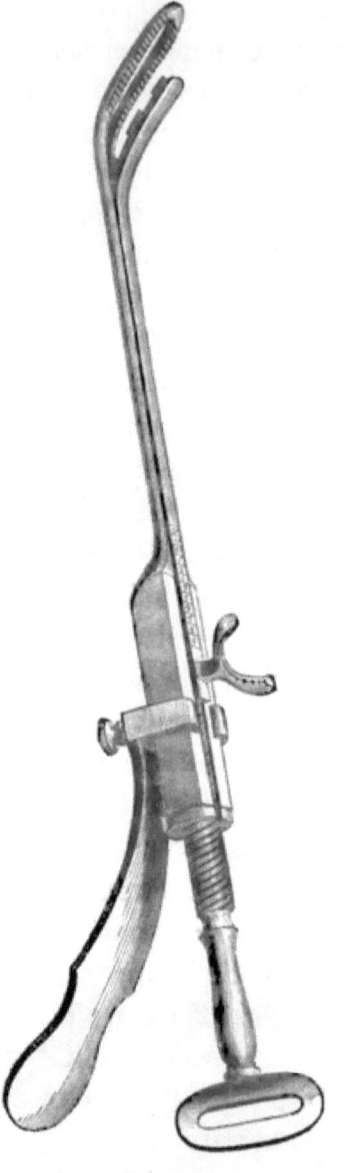

CHAPTER IX.

DISEASES OF THE URINARY AND GENERATIVE ORGANS IN THE MALE.

I. RUPTURED URETHRA IN THE MALE.

OF all the miserable accidents that can happen to a young lad, that of falling astride across a bar and rupturing his urethra, is the one which, not being immediately fatal, is attended with the most disastrous results. Its effects, in a bad case, are never recovered from. Stricture follows, with its attendant miseries, and these abide more or less through life. The mode in which the accident happens is usually this:—A boy is climbing over railings, or walking along a ladder, or playing some such prank, when his foot slips, and he falls with his legs apart upon a hard narrow body, striking the perinæum. The immediate consequences are more or less effusion of blood, pain and swelling in the injured part, and bleeding from the urethra. If the urethra be entirely divided, there is inability to pass the urine; if it be only partially ruptured, the urine may be passed, but it will be accompanied with a large quantity of blood. There can be no difficulty in diagnosing this accident; and when the surgeon has determined its nature, upon the promptitude of his action will depend the well-being of his patient. An attempt should be cautiously made to pass a catheter, and if the urethra be not completely divided, this attempt may perhaps prove successful; a result devoutly to be wished. The advantage to the patient cannot be over-estimated. If the catheter can be introduced, it is on no account to be withdrawn for several days. The urine will pass off by that channel; or if there be infiltration by the side of it, and an abscess accordingly form in the perinæum, it should be opened when suppuration is established. In these, the most favourable cases, the formation of such an abscess is not attended with any very great amount of inconvenience.

If, on the contrary, the catheter cannot be introduced easily, on no account should any forcible attempt be made, as most probably what has happened is, that the urethra is completely divided, and the end retracted, under which circumstances the passage of the catheter into the bladder is a mere impossibility. The young patient should be at once chloroformized, placed in the same position as for lithotomy, and the surgeon, exactly fixing on the mesian line (which, owing to the swelling and the effusion of blood, is difficult to recognise), should plunge a double-edged knife into the perinæum, in the direction of the injured part of the urethra. Having laid open the parts, and fortunately hit the vesical portion of the urethra, a gum-elastic catheter may be passed into the bladder through the wound, and fixed by tapes around the pelvis. This should be retained for four or five days, and then withdrawn. The foregoing is the most orthodox plan of treatment, but it is quite as good a plan simply to lay open the perinæum in the direction of the urethra, taking care to cut deep enough, and leave the urine to pass out through the wound. In the course of a fortnight the urine will begin to pass through the natural channel, and the wound in the perinæum to heal. The great difficulty in micturition does not occur for some weeks or months afterwards, when contraction takes place in the tissues forming the new urethra. When this symptom appears, a catheter should be passed every two or three months, but the relief given is slight, and the prospect of permanent benefit unhappily very small.

It is usually recommended to pass as large-sized a catheter as possible at the time of the injury; but according to my experience in the matter, a small one is more likely to be successful.

Passing the Catheter.—This, even in the adult, delicate operation, demands, in the case of children, the greatest precaution. Every one knows the unadvisability of forcible catheterism; in the case of children, it is positively unjustifiable. The lightest hand and most delicate manipulation will always suffice, if the catheter be of proper form and size. The shape of the instrument which I employ, is shown in the accompanying woodcut, and it is one that is passed with much greater facility than any other. It will be noticed, that it describes a considerably smaller curve than the usual adult catheter. Bleeding should never follow its use on a child; if hæmorrhage occur, the surgeon may have reason to suspect that a wound has been made in some portion of the urethra. The mucous membrane of this canal in the child is much more delicate than in the adult; and, inasmuch as the operation should be performed under the influence of chloroform, double care is necessary; since not only are the indications of pain wanting, but the parts themselves are relaxed and offer less resistance.

II. CALCULUS IN THE URETHRA.

The causes of retention of urine in the adult are very various; in the child, with the exception of accidental injury as above described, there is only one— the impaction of a calculus in the urethra. Congenital phymosis might be supposed likely to give rise to it, but this, however completely it may close the opening of the prepuce, always allows the urine to dribble away. When, therefore, a child is brought to us with complete retention of urine, we may be pretty sure of the nature of the affection, and be prepared to treat it accordingly. In these cases, the inability to pass water is the first symptom; there is no history of stone. Generally, when the surgeon first sees the case, the child is suffering great distress, and the distended bladder may be distinctly felt above the pubes and through the rectum.

The *Treatment* of retention of urine is, first of all, to place the patient under the influence of chloroform.

Case.—A. B., aged two years, was admitted into Guy's Hospital, in January, 1860. The bladder was distended as high as the umbilicus. Chloroform was administered; as it took effect, a small lithic acid calculus was shot from the urethra, some little distance into the bed. The urine followed in a full stream, and the child left the hospital in about an hour.

Only actual experience can enable us fully to appreciate the value of chloroform, and the instantaneous relief it gives, not only in some instances of retention of urine, but in numerous other affections of children. It is not, however, always the case that the calculus is expelled on the administration of the drug, and if the retention continues, a catheter must be passed. In doing this, the calculus may be distinctly felt, and accidentally pushed back into the bladder, where its presence may or may not be afterwards detected by the sound. In the following case, it could not be detected, and it is fair to presume that it passed away unobserved.

Case.—J. B., aged four years, was admitted into Guy's Hospital on the 11th January, 1860. On the 9th, he had complained of pain in the penis. On the morning of the 10th, he had sudden difficulty in micturition, which continued until his admission on the following morning. The bladder was then found distended up to the umbilicus. Chloroform was given, and the urine not being expelled, I passed a silver catheter. In doing this, a stone was felt in the urethra, and was pushed back into the bladder. He then passed part of his urine, the remainder being drawn off by catheter. He remained in the hospital until the 27th, being sounded by Mr. Hilton several times without detecting anything. The father of this patient, when twelve years old, was cut for stone by Mr. Cock.

In the following case, the calculus, though not observed at the time of its passage, was afterwards found.

Case.—J. B., aged four years, was admitted into Guy's Hospital, in May, 1855, with retention of urine. The boy was suffering great agony. He was put under the influence of chloroform, and the catheter was passed. A stone was not felt, though no doubt was entertained that one existed. The urine was passed involuntarily every night, and on the fifth day a small calculus was found in the bed, by the nurse, who had been directed to watch for it.

Retention of urine, however, is not a necessary symptom of calculus in the urethra. On the contrary, there may be incontinence of urine; and, indeed, these two symptoms, as it were, divide the cases of calculus in the urethra between them; if there be not perfect retention, there is always incontinence. This incontinence may sometimes last for a long time before its cause is discovered, and it may be attended with various other symptoms, such as pain in passing water, and occasional sudden stoppages of the stream. The symptoms of calculus in the urethra, when attended with incontinence of urine, indeed, are very similar to those of stone in the bladder, but less severe. In all cases where any symptoms of this kind are present, the surgeon should pass a sound. This is a proceeding which should never be omitted. If there be a calculus in the urethra, its presence will be felt at some portion or other of the canal. Should there be a very contracted prepuce, the stone may be lodged there, in which case it may be sometimes felt by the fingers. I was called by my friend Mr. Greenwood, of Horselydown, to see a little boy, with symptoms of calculus in the urethra. In this case, I found three small calculi lodged in a long prepuce. The explanation of these cases I take to be that, owing to the small opening of the prepuce, the urine accumulates within it, and gradually forms the calculi where they are found; the secretion of the glandulæ Tysoni constituting a nucleus.

If not contained within the prepuce, the calculus may be found at any other part of the urethra, but most frequently about the bulb. Various plans may be adopted for its removal, according to the part of the canal at which it is lodged. If just within the meatus, the orifice may be slit open, and the calculus abstracted. If situated in any other part of the urethra, a long narrow pair of forceps has been recommended to be passed down and the stone seized. This, however, is an unsatisfactory proceeding in children, and I have never adopted it, believing that it is fraught with less danger and injury to the canal to pass a grooved staff down to the stone, and cut into the urethra in the mesian line from the perinæum, or from the under part of the penis, either in front of or behind the scrotum. I must advert here to a maxim which is inculcated, but which I am satisfied is incorrect—viz., that the urethra should never be opened in front of the scrotum. I have opened it there myself, and seen the same thing done by others, and the wound has healed quite as rapidly as when the incision was made in the perinæum.

When the urethra is opened, the stone is to be immediately seized by the forceps, and extracted. Sometimes it is best dislodged by a curette passed behind it. The after treatment is simply to apply lint dipped in warm water to the wound, taking care to tie any vessels that may bleed, and to keep the child clean and *quiet*—that is, well amused. The urine passes partly through the wound for a few days or weeks, the time depending on the health and vigour of the patient.

In the following case the incision was made in front of the scrotum:—

Case 1.—C. F., aged ten years, was admitted into the Infirmary for Children in August, 1858. He was a delicate child: complained of pain in passing his water, and for the last five months there had been partial incontinence. I immediately introduced a sound, and detected a stone in the urethra, in front of the scrotum. Chloroform was given, and I at once cut down upon and removed the stone. The wound healed rapidly; he had not a bad symptom, and went out quite well in thirteen days. This operation, however, is not always so satisfactory in its progress.

Case 2.—J. F., aged nine years, was admitted into Guy's Hospital in April, 1858, under Mr. Birkett's care. The history given was, that a stone had been removed from the urethra by the forceps three years previously. Three months before admission he complained of pain in the penis on passing his water, which also sometimes passed involuntarily. Medical treatment had been adopted, but a sound had not been introduced. When admitted, a calculus was found in the urethra, in front of the scrotum. Chloroform was administered, and the stone, which was of large size for its position, was removed through an incision in front of the scrotum. The hæmorrhage being rather more than usual, cold and pressure were applied. It is conjectured that the latter, obstructing the flow of urine, caused its extravasation into the scrotum. Incisions were required for the relief of this symptom, but the boy ultimately did well, and was discharged cured in a month.

The following is an example of an ordinary case of this affection. The calculus had apparently been lodged in the urethra for some time.

Case.—S. G., aged seven years, came under my care at the Surrey Dispensary in February, 1850: a pale, cachectic-looking boy, though not in actual ill-health. The mother stated that she found, after breeching him twelve months ago, that his clothes were always wet. The urine is now continually passing away from him, and he has a little pain in micturition. The stream also sometimes stops suddenly; there has never been any blood. He does not wet the bed, but rises six or eight times in the night to pass his urine. He has been several times under medical treatment during the year, without any permanent advantage. I passed a sound down the urethra, and struck a stone lying at the bulb. I at once cut down upon and extracted it. The recovery was rather tedious, and it was a month before he was perfectly well.

The calculus in this case was of a peculiar form. There was a distinct groove on its surface, in the direction of its longer axis, as if the urine in its passage had worn a channel for itself. So marked was this appearance, that when the stone was shown to the mother she refused to believe it had been removed from the child, and persisted that it was a coffee berry, which had been secreted in my pocket to impose upon her. I conceive that the calculus in this case had become sacculated in the urethra, without destroying the integrity of its lining membrane.

III. EXTRAVASATION OF URINE.

In relation to this subject two general principles may be laid down: First, that swelling of or around the urinary organs, in boys, is in almost every case caused by extravasation of urine; and secondly, that extravasation of urine, if not due to accident, is invariably caused by calculus in the urethra.

I have said that swelling about the urinary organs is in *almost* every case due to extravasation of urine, because exceptions do occur, though very rarely.

The following case is the only instance I have ever seen in which there existed such swelling without extravasation, and in it the peculiar slight redness of the skin, which soon appears in extravasation of urine, was absent, and the swelling was principally confined to the pubes, the perinæum and scrotum not being involved.

Case.—A boy, aged seven weeks, was brought to me at the out-patient room at Guy's Hospital, in the year 1857, with swelling and hardness over the pubes, and slight constitutional irritation. The case was watched for two or three days, the history not being sufficiently clear to warrant an incision. The child evidently became worse, it was difficult to say from what cause, and on the fourth day he seemed to be sinking. Thinking it possible that there might be some extravasation of urine, an incision was made into the swelling. Troublesome hæmorrhage occurred, and in the course of the evening the child died. On the post-mortem examination, there appeared nothing but an œdematous condition of the tissues. No calculus was found.

When a case of extravasation of urine is brought before us, there is more or less swelling of the perinæum, scrotum, penis, and lower part of the abdomen, the amount depending on the time that has elapsed since the urethra gave way. This swelling, as is well known, is limited posteriorly by the union of the superficial with the deep perinæal fascia, but there is nothing to limit its extent upwards on the abdomen. I have seen it, as in a case which I shall hereafter mention, extending as high as the scrobiculus cordis. In that case there was also a peculiar yellowish appearance of the skin, which I have never met with since, but which I should easily recognise again. The constitutional disturbance varies very much; I once saw a boy almost entirely prostrated, as if from uræmic poisoning, at a comparatively early period; at other times the children seem to suffer very little. These differences, I believe, depend chiefly upon the rapidity with which the extravasation takes place. If it go on slowly, the swelling will gradually increase, and the tissues will seem almost to accommodate themselves to the irritation; so that the obvious symptoms will very inadequately represent the extremity of the danger. These slow and insidious cases are among the worst; they have often been going on for several days before the child is brought to the surgeon, and almost irreparable injury is done before any steps are taken to avert it.

The mother tells us—that is, if we closely question her, as we should always do—that there has been a diminution of the flow of urine, and a difficulty in passing it, and that sometimes it has suddenly ceased to flow. Complete retention of urine is not a frequent symptom; but, in children, any diminution in the quantity of urine passed, especially if attended with swelling or pain in the perinæum, demands our serious consideration. There is no question, that in them, as in adults, extravasation of urine, which may in time reach a fatal extent, can be slowly taking place, although a certain quantity of urine escapes by the natural passage. Adult patients seem so unconscious of the amount of damage which may be going on in the perinæum, if urine to any moderate amount passes through the meatus, that we are frequently baffled in our attempt to get any account of the condition of the urinary organs, by the assurance on the part of the patients that they are passing their water. If this is the case in adults, how much more likely is it that in children the first symptoms of this disease should be overlooked or treated with indifference, unless, as sometimes fortunately happens for the little patient, the symptoms of stone in the bladder have been rather urgent for some weeks previously, or the constitutional disturbance happen to be severe.

The age at which this accident generally takes place in children, is from two to seven years; after that period, it is much less frequent; the calculus, if one be present, does not appear easily to escape into the urethra, and lithotomy accordingly is required.

If the previous account of the cause of extravasation of urine in children be correct—that is, if apart from accident, calculus in the urethra be the only means by which it can be produced—the treatment becomes very clear. The child being placed under chloroform, the first thing to be done is to pass a grooved sound into the urethra, to ascertain if a stone can be detected in that canal. If we are fortunate enough to detect one, an incision should be made through the swollen parts into the urethra, and the stone removed. If, on the contrary, no stone can be felt, then free incisions must be made into the infiltrated parts, to permit the escape of the urine and subsequent sloughs. A quantity of fluid passes away, having an urinous odour, and sometimes there is a good deal of bleeding. It is necessary to be very cautious that this bleeding do not proceed to any great extent. I have seen cases in which fatal hæmorrhage occurred. The incisions should be made as near as possible to the mesian line.

If we cannot discover a stone, we may nevertheless rest sure that one is present, and has escaped into the areolar tissue: the nurse should be directed to be on the watch for it, if only to verify our diagnosis. It may be weeks before it will be found. When the bleeding has ceased, hot poppy fomentations should be applied; and if there be swelling of the prepuce, it may be slit up to allow a free escape of urine by the urethra. Stimulants and nourishment should be liberally administered. Wine is indispensable; often brandy will be required. An opiate should be given immediately after the operation, and repeated whenever the child is restless. Bark and acid, or quinine, are the best internal remedies.

A question of practice should here be adverted to: Is it advisable to keep a catheter in the bladder? My own impression is, decidedly not; as, although it prevents the urine from escaping through the opening in the urethra, and so giving rise to fresh extravasation, yet should the catheter become blocked up from any cause, it may produce the very results we desire to avoid. If, however, the catheter be used, great care should be taken, first that it is sufficiently large to fill the urethra; and secondly, that it does not become obstructed, which involves frequent removal; the risk thus incurred is evident.

For external dressing, lint, dipped in warm water, is the best. This should be frequently changed; and, therefore, need not be covered with the thin gutta percha, unless the patient is obliged to be left for a longer interval than an hour. The discharge is profuse and sloughy; large shreds of dead tissue will have to be gently removed from the various incisions, and the destruction of the surface is often very great. I have seen the corpora cavernosa and both testicles exposed. At other times, with less destruction of the soft parts, the child appears to have sunk, without power to throw off the sloughs. In these cases, it is of little service to apply stimulants to the wound, unless the granulations be very flabby; the repair must be effected by the rallying of the vital powers within. Thorough good nursing and cleanliness are of the utmost importance. Cleanliness especially is to be emphatically enforced; constant but gentle cleansing of the parts surrounding the wound, is essential to prevent their becoming excoriated; and the little patients may be carefully dipped in warm baths with advantage.

In spite of our best efforts, it will happen now and then that a little fresh extravasation will occur, more particularly if the stone have not passed away. It is useless, in these cases, to attempt to find it; and annoying as the circumstance is, we must confine our efforts to supporting the general powers. Unhappily, unless these are vigorous, the renewed extravasation is generally fatal. Of course the surgeon's great object in every case will be to extract the stone at the time of the first incision; if this can be done, the subsequent progress of the case is simple, but even then, if the extravasation have been great, death will most probably ensue.

Sometimes, instead of passing by the natural channel, the urine flows entirely through some of the wounds. If this continue beyond a fortnight or three weeks, it is advisable to pass a bougie or small catheter (of the shape before recommended) through the urethra. This should be cautiously done, anything like force being carefully avoided; it should be repeated once or twice a week until the urine escapes entirely through the urethra.

The following cases illustrate several of the points which have been referred to. They all came under my observation in Guy's Hospital.

Case 1.—A delicate boy, aged two years, was admitted in May, 1854. The day previous he had retention of urine, which was quickly followed by extravasation, involving the perinæum and scrotum; incisions were made into these parts, and a small calculus came away. On the next day a catheter was passed and retained in the bladder; this became blocked up from some cause, and extravasation again occurred, involving the lower part of the abdomen. Fresh incisions were made, but from the amount of sloughing and constitutional irritation, the child died on the ninth day.

Post-mortem Examination.—Peritoneum covered with buttery lymph. In the scrotum there was a sloughing wound, which passed backwards to the base of the bladder, anteriorly communicated with another wound near the glans penis; the whole of the cellular tissue of these parts being involved. There was also sloughing external to the prostate, reaching to the left side within one eighth of an inch of the peritoneal fold; the incisions from the perinæum extended into the prostatic portion of the urethra. All the other organs healthy.

Case 2.—A boy, aged five years, was admitted in February, 1856. On the day previous he had difficulty in passing his water, and after a few hours it ceased to flow; extravasation of urine occurred, involving the perinæum and scrotum, which were considerably distended. Incisions were made into the parts, but the extravasation increased, and shortly a calculus made its exit. There was a peculiar livid appearance upon the abdomen, the veins were distended, and skin of a yellow colour, the extravasation having extended as high as the scrobiculus cordis. From the immense amount of sloughing, the child sank exhausted on the twenty-eighth day after admission.

Post-mortem Examination.—The inflammation and sloughing of the cellular tissue had involved the abdominal muscles; these were infiltrated with a purulent secretion of a green colour, and in places were commencing to slough. The peritoneum lining these had escaped the contagion in a most remarkable manner, there being not the slightest trace of peritonitis. The sloughing wound in the perinæum and scrotum laid open a considerable part of the urethra; nearly the whole of the spongy portion, excepting that in the glans, was exposed: thus there remained only the anterior surface of that canal intact.

Case 3.—A delicate boy, aged four years, was admitted May, 1856, with symptoms of calculus in the bladder, and a stone was detected. He was brought into the operating theatre, but no stone could be felt, consequently nothing was done. Five days after this, extravasation was found to have taken place, involving the scrotum, penis, and lower part of the abdomen; there was also great prostration. A sound was readily passed into the bladder, and the perinæum and urethra laid open; no stone could be detected. The child never rallied, and died in twelve hours.

Post-mortem Examination.—The urethra opposite the scrotum was ulcerated, extravasation having occurred from this spot; and in a sloughing cavity lay a small stone.

Case 4.—A stout, hearty little boy, aged twenty-nine months, was admitted September, 1856, with swelling of the perinæum, scrotum, and penis, from extravasation of urine; considerable difficulty in passing water had been experienced, but not complete retention. A gum-elastic catheter was passed to empty the bladder, and incisions were made into the swollen parts; fresh extravasation occurred at intervals, for seven weeks, accompanied necessarily with sloughing; at the end of that time a small calculus dropped out from one of the several openings which had been made. He never afterwards had a bad symptom, and quickly recovered.

Case 5.—A delicate boy, aged three years, was admitted in May, 1857. He had pain in passing water for three or four weeks. Nine days before admission had partial retention; the scrotum, penis, and lower part of abdomen began to swell, and became immensely distended; sloughing commenced in various parts, and very little urine passed. On admission, he was almost in a dying condition. Chloroform was administered, and free incisions were made into all the parts. The phymosed prepuce was slit up, and a grooved director passed into the urethra; a stone, which was felt in front of the scrotum, was cut down upon and removed. He quickly rallied, with brandy and good nourishment. The whole of the sloughing parts were thrown off, and the wound soon began to granulate. On the eighteenth day after admission, the urine not having passed by the urethra, an attempt was made to introduce a small catheter into the bladder, but unsuccessfully; on the twenty-fifth day a bougie was readily passed. This was one of the most remarkable cases of recovery after extravasation of urine which I have ever seen.

Of these five cases death occurred in three; and yet, in one of those three, the calculus came away directly. But it must be remembered that in this instance a catheter was passed, which became obstructed, and, therefore, gave rise to the same conditions as if the stone had not been removed. It is true that in

Case No. 4 the calculus was seven weeks before it came away, and yet the patient recovered; but it should be borne in mind that this boy was a remarkably healthy and strong little fellow; and though he lived so long, and ultimately recovered, there was scarcely a day during that time in which there was not the greatest anxiety as to the termination of the case. It was not until the stone dropped out that all danger was at an end, and that he commenced to do well. The last case was undoubtedly the worst—nay, it appeared almost hopeless when first seen, and I can only attribute the successful issue of it to the removal of the stone immediately the boy came under notice.

I think we have evidence, therefore, that any delay in removing the stone increases the prospect of a fatal termination, or, at all events, renders the little patient less able to withstand the great constitutional disturbance that must necessarily arise.

The practical directions to be gathered from these cases are as follows:—

1. Attend to the earliest symptom, viz., retention of urine in children.

2. Incise freely the parts into which the urine has been extravasated.

3. Search for and remove the calculus immediately.

4. Do not pass a catheter.

IV. INCONTINENCE OF URINE.

If in children, extravasation of urine arises from but one cause, incontinence of urine may arise from many, and these very different in their nature. There are, indeed, few affections of the urinary organs in children, of which incontinence is not more or less a symptom; in addition to which it arises apparently from undefined affections of the nervous system, and it is said to be caused by irritation of the intestinal canal. Unfortunately, when from any cause, incontinence of urine has become a habit during childhood, it is apt to continue till towards the period of puberty. Almost always the boy is very much ashamed of the fact, so that it may be regarded as a true disease, not dependent on want of exertion of the will. Incontinence of urine is almost an unfailing symptom of stone, whether in the bladder or the urethra, as already observed, also of growths within the bladder; and besides these, there are some trivial affections of the urinary organs, such as a long prepuce, &c., which may give rise to symptoms of incontinence, either by causing irritation of the glans, and so disordering the action of the bladder, or by the mere mechanical retention of the urine at the outlet. This affection has also been attributed to the existence of lithic acid in the urine; but I suspect that these cases, which are often treated with temporary success by the physician, eventually find their way to the surgeon, with other and more marked symptoms of calculus.

In the worst cases, the little patients, when playing about, are suddenly seized with an irresistible need to pass their urine, and if the means for their relief be not instantly afforded, it is passed against their will. In consequence of the frequency with which it escapes, the quantity voided at each time is necessarily small. At night, the same condition continues; the urine being passed not once only, but several times; and it is useless to attempt to prevent this by waking them at stated periods. In other cases, the affection is confined to the night; and when this form of the disease occurs in older children, some benefit may be derived from the practice of waking them. But it is a curious fact, that the patients in whom incontinence occurs both by day and night, are the most amenable to treatment.

When not dependent upon any disease of the urethra or bladder, incontinence occurs generally, though not exclusively, in weak and sickly children, and particularly in those resident in towns. As their general condition improves, the affection diminishes; and, as I have often seen, tends to return, if their health again deteriorates. In all cases of incontinence of urine, however, an accurate diagnosis is of the utmost importance; and the more since it is to be made by a process of exclusion. Always, the assumption should be that there exists a calculus, or other mechanical cause, until its absence has been demonstrated by careful exploration.

The *Treatment* of this affection depends upon its cause. The special causes, such as calculus, long prepuce, &c., will require their appropriate operative procedure. If the general health be impaired, various remedies are indicated according to the circumstances; among them, tonics hold an important place. Iron is useful in its way; but I must offer my protest against the indiscriminate manner in which this remedy is sometimes employed in diseases of the urinary organs. Change of air, healthy exercise, mild and nutritious diet, and cold sponging, are the agents on which chief reliance is to be placed. I have known many children who suffered in this way during residence in town, and whom a few days in the country quite relieved. Belladonna and strychnine have been recommended, as also tincture of cantharides and nitrate of potash. If these opposite remedies really do good, it would be difficult to assign a rational pathology to the disease. I have not, myself, found reason to rely upon any of them; but I have no doubt that the surgeons who employed them, did not fail, also, to enjoin dietetic and other hygienic rules. In nocturnal incontinence, it has been recommended to apply a blister over the sacrum, to prevent the patients from lying on their backs; as by this means, it is supposed that the urine is prevented from irritating the most sensitive part of the bladder, the trigone. This I believe to be a fallacy; and not less anatomically than physiologically. I have never seen a case which I could distinctly trace to the presence of worms in the rectum, and cannot, therefore, speak of the value of vermifuge remedies. Occasionally, mild aperients are beneficial, but I believe this is through their effect on the general system.

After what has been said, it is hardly necessary to add that any corporeal punishment in these cases is most unjustifiable and cruel. It is the surgeon's place to be the child's protector. The shame it

DISEASES OF THE GENERATIVE ORGANS.   159

suffers, if punishment were needed, is punishment enough. This shame, or the fear of being punished, will sometimes induce the poor child to use mechanical means to prevent the flow of urine, and these may have serious results, as in the following case:—
D. C., aged eight years, was admitted into Guy's Hospital, under Mr. Birkett's care, in January, 1854. He had been severely chastised for wetting the bed,

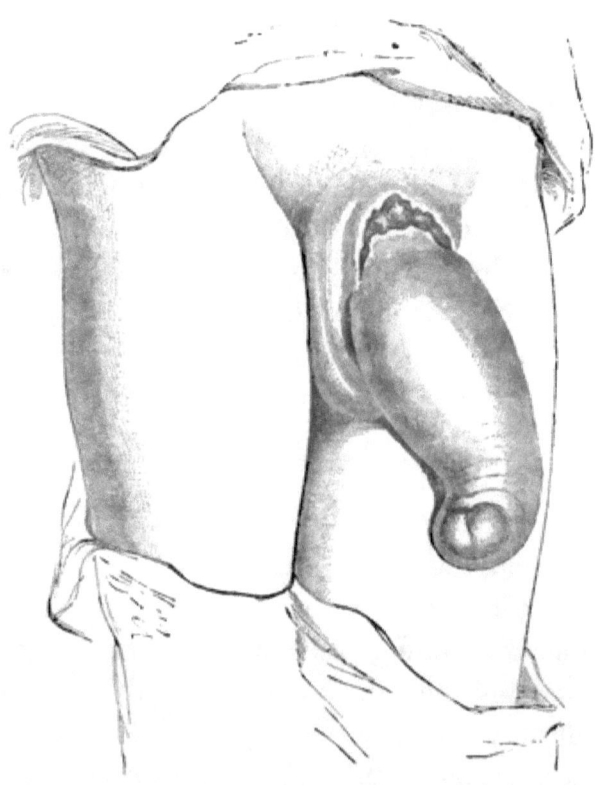

and ten days before admission he had tied a piece of string round the root of his penis on going to bed.

On admission, there was a deep ulceration all round at the root of the penis, great swelling, and partial retention of urine. The string was cut away, incisions were made to relieve the œdema, and the child recovered without permanent damage. The foregoing woodcut represents the condition of the penis soon after admission.

Impressions upon the mind, however, may be useful to the patient. Bread pills, for instance, seem at times to give an imaginative stimulus, that helps to break the chain of habit. I have also frequently found the passage of a sound of great benefit, partly I think in a similar way, as there is a certain amount of pain attending its introduction. It seems, also, to have the effect of diminishing the irritability of the neck of the bladder, and there is an additional recommendation in this practice, as now and then a calculus has been struck when it has been least expected.

V. POLYPUS OF THE BLADDER.

Whilst speaking of diseases of the bladder, I cannot omit the mention of a very rare case which occurred in Guy's Hospital, under Mr. Hilton's care, about two years ago:—

Case.—W. T., aged two years, was admitted in Aug. 1858, with a swelling in the abdomen, apparently connected with the bladder, and attended with obscure urinary symptoms. Soon after admission the patient had an attack of acute peritonitis, and in the course of a month he died. *Post-mortem Examination.*—On opening the abdomen, the tumour, felt during life above the pubes, was found to be the enlarged and thickened bladder, pressed forward by a growth below. This sprang from the posterior surface of the bladder, and was almost the size of an orange. The rectum passed over it, pressed upon, but not implicated in the growth. The tumour was circumscribed, and did not involve any of the surrounding structures, so that when the bladder was taken out, all the disease was removed. The coats of the bladder were much thickened, the muscular coat especially being hypertrophied; and within it was a mass of polypi springing from its neck, and surrounding the orifice of the urethra. These had slender pedicles, were united at their origins, and formed distinct polypoid growths, the whole mass being about the size of a pigeon's egg. The post-mortem was conducted by Dr. Wilks, to whom I am also indebted for the following account of the microscopical characters of the tumours. The polypoid growths, on section, were white and firm, and did not emit any juice on pressure. The microscope showed the structure to consist of an interwoven simple fibrous tissue, with small oval nuclei. The mass, external to the bladder, was of the same character, but was softer, partly from a process of degeneration which seemed to be going on throughout it.

The growth had exerted pressure on the ureters, both of which were much dilated, as were also the pelves of the kidneys. The right kidney was suppurating throughout its whole substance, and this to a much greater extent than is usually seen in pyelitis. The organ was enlarged to twice its natural size, the capsule was implicated, and the pus escaped by numerous points into the peritonæum.

CHAPTER X.

CALCULUS IN THE BLADDER.

THE subject of calculus in the bladder in children must ever be of great interest to the surgeon, and remembering the happy results of a successful operation, it can hardly be otherwise than one of the most captivating. Any one who has walked round the wards of a hospital, as I have done at the Infirmary for Children, with five of these little fellows suffering in various ways from this painful malady, must be devoid of feeling if he does not reflect, with a genuine pleasure, upon the results which are almost sure to attend the exercise of his art; and that in the course of a very few weeks he may have the satisfaction of seeing these same children well and playing happily. It is not my intention to enter into a long history of calculus, its causes, relative prevalence in different localities, &c., inasmuch as I have to deal only with a certain section of the community, and one, perhaps, less influenced by local conditions than any other. But I may refer, in passing, to the remarkable frequency of this affection in children as compared with adults. My impression is, that it may even exist at the time of birth. I have not had the opportunity of making examinations of still-born children, but in two or three cases I have seen undoubted symptoms of calculus, pain in passing urine, and a little blood, very soon after birth. These symptoms continued, in one case, until the boy was two years and a half old, when he was cut for stone. The frequency of calculus among the children of the poorer population of London, and its comparative absence among the richer, are, I believe, to be attributed to some difference in diet, and probably to the want of animal food. This, however, is a subject upon which our evidence is as yet incomplete.

With these few remarks, I shall leave the general question of calculus, and shall proceed to describe the symptoms which it presents in early life. I would first observe that children, especially young children, being unable to express their feelings accurately, symptoms may exist for weeks or months without exciting particular attention; or, if noticed, they may be so trifling as not to induce the mother to seek advice, more especially as the most marked is one that is very common. My readers will have anticipated that symptom to which I especially allude: it is one which is always present, and if coupled with a deposit of mucus in the urine may be regarded as pathognomonic; I mean incontinence of urine. When a child with stone voids its urine, a little shivering occurs; it seems to have a cold creeping sensation, and will often complain of great pain, referred to the end of the penis, and which the child tries to relieve by squeezing it, and pulling it out. Thus is produced an elongated state of the prepuce; but this is not to be relied on as a special symptom, since it may equally exist in congenital phymosis, and may then even give rise to symptoms resembling those of stone; viz., an apparent incontinence of urine, with uneasiness at the end of the penis.

Occasionally a small quantity of blood will pass. Many of these patients, however, never have this as a symptom at all; but, on the other hand, in one patient that came under my care, it was the only symptom. A sudden and painful stoppage of the flow of urine occurs in some cases. If the urine be passed into an utensil, there is always a deposit as it cools, either of mucus or of pus. Young boys with stone seldom suffer as much from jumping and running about as adults, but still they complain a little. The peculiar pains about the pelvis and thighs met with in adults, I have not observed in children; but they have very frequently a fixed pain deeply seated, just over the pubes. In the act of micturition they very frequently have prolapsus ani, in consequence of the prolonged and repeated straining. In one case, that of a boy aged three years, the chief symptom was irritability of the rectum, and pain in passing his motions. There was not any disturbance of the urinary organs. Even the prepuce did not at all resemble that of a boy with stone. These symptoms occurring together, or any of them individually, may induce the surgeon to believe that the child has a calculus, but there is of course only one perfectly diagnostic sign, that obtained by the use of the sound.

All the symptoms of stone are much more severe during the summer than the winter months; this I am disposed to attribute to the greater concentration of the urine during the warmer weather. The irritation it produces is inversely as its dilution. Thus it happens that the majority of cases are brought for treatment in the summer.

Sounding for Stone.—The sound employed should be of the shape shown in the accompanying woodcut. It will be noticed that it has a bulbous end, which is accordingly larger than the stem. In the instrument depicted the bulb is of size No. 3. It may also be plated with silver. And the same precautions should be taken in the use of this instrument as in that of the catheter. Under no conditions should force ever be employed. The patient may either be subjected to the influence of chloroform, and allowed to lie on a bed or couch, or he may be held in the position (a very convenient one), shown in the woodcut opposite; the surgeon then only requires the assistance of a good nurse. The sound being well oiled, should be allowed to glide almost by its own weight into the bladder, when if the stone be of any size, it will be struck immediately, almost always giving a sensation as if it were at the lower end of the urethra. When the stone is very small, it is desirable that there should be water in the bladder; but if a moderate quantity be present, even a small stone is easily detected. The sharpness of the curved portion of the sound permits its free rotation in the bladder.

If the stone be not struck at once, a light jerk of the hand will often cause it to fall against the instrument.

Having ascertained the presence of the calculus, the question is, how should it be got rid of? Two modes of proceeding present themselves—lithotomy and lithotrity. In the latter of these operations, so far as boys are concerned, I have no experience; and when we consider the requirements essential to a successful case of lithotrity: viz. a bladder holding a fair quantity of urine, and a urethra large enough not only to admit the passage of the fragments of the stone when broken, but also of a good-sized instrument; and when we also remember the remarkable success which attends the operation for lithotomy in these young patients; we can hardly feel disposed, even though recommended by Civiale, to perform lithotrity in boys. These are the reasons for which I have in my own practice rejected it; and yet I have once or twice seen so small a calculus extracted by lithotomy, that I have thought a lithotrite of very small size might perhaps have been used. It is not, however, my intention to describe the operation in question, especially as the method of procedure is the same in children as in adults. I accordingly proceed to notice

The Operation of Lithotomy.—Two modes of performing lithotomy are in use; the ordinary lateral operation, and the method called Allarton's, as having been re-introduced by a gentleman of that name. The latter plan consists in making the incision in the mesian line of the perinæum, instead of obliquely in the ischio-rectal fossa, cutting into the urethra, in which a grooved staff has been placed, and then with one finger in the rectum (to avoid wounding that viscus), carrying the knife carefully along the groove towards the bladder. The knife being withdrawn, the finger is passed into the bladder to dilate its neck, and the forceps being introduced the stone is extracted. This operation is applicable only when the stone is small; its recommendation is that there is likely to be less hæmorrhage than in the ordinary plan. I have seen this operation performed with very slight loss of blood; but on the other hand, I have witnessed cases in which the bleeding was very considerable, quite as great as in the other method. It appears to me that there are no particular grounds for performing an operation of this kind, when the usual oblique incision is as successful in its results in children as any serious operation can be. It should be remembered, also, that the mesian incision is not available for the extraction of large calculi, and every surgeon must have experienced the difficulty of gauging the size of a stone prior to its extraction. The plan itself has been tried before, and abandoned for the simpler lateral proceeding, doubtless not without reason. Though not averse to novelties in operating, I should hesitate to discard a well-tried and satisfactory proceeding for a plan, the advantages of which are in every case doubtful, and which is not available for all. It, therefore, remains for me to describe, which I will do in detail, the steps connected with the lateral operation of lithotomy.

A previous systematic preparation of the patient, on the part of the surgeon, is unnecessary. But, of course, any derangement of the intestinal canal, or general system, should be rectified. As regards any organic diseases, of the viscera, spine, &c., that may be present, they should not deter us from the operation. It would be right to relieve the patient from the urgent symptoms connected with the presence of stone in the bladder, even under conditions the most unfavourable to his permanent recovery. As for disease of the kidney, its diagnosis is so beset with difficulties in cases of stone, that its existence need hardly be made matter of consideration: the urine may present all the symptoms of kidney disease, be albuminous, or loaded with pus or blood, merely from the irritation of the bladder.

On the evening previous to the operation, I invariably give these children a calomel and jalap purge, with the view of thoroughly emptying their bowels; and if they have not been well relieved, an hour before the operation, I order an injection of warm water. Five or six assistants are required:—one to give the chloroform, one to hold the staff, two to tie up or hold the patient, and keep him still, and one to hand the necessary instruments to the operator. A stout, strong table is necessary, as likewise a chair, with about six inches cut off its legs, to enable the operator to sit, so that his neck shall be about level with the perinæum of the patient.

The child being placed in a blanket upon the operating table, and chloroform given, the assistants should proceed to tie up his hands and feet. Sometimes, in the case of very young boys, I dispense with the tying, but then great care must be taken by the assistants to keep the pelvis fixed. The surgeon, standing between the legs of the patient, should next introduce into the bladder a staff that fits the urethra rather tightly, great care being taken not to wound the passage. A little difficulty is sometimes met with at the triangular ligament, and if any force be used, it is possible that the staff may slip anywhere but into the bladder. The surgeon should not grudge a minute or two to perform this part of the operation perfectly. Many a young operator has become embarrassed, by finding more difficulty than he expected in this stage of the proceedings. The straight staff, which the profession owes to Mr. Key, is the one which I always use. The point of it should be brought to rest either upon the stone, or just within the neck of the bladder. It should then be given into the hand of an assistant, and held firmly in its position until taken from him by the operator.

If the stone cannot be struck, the staff must be withdrawn, and the sound should be used, with the view of exploring the bladder, and assuring ourselves of the existence of the stone. It is a wholesome rule, on no account to be disregarded, that the calculus should be struck immediately before the operation. If the surgeon fail to detect it, the patient should be returned to his bed, and the operation postponed till another day. It is difficult to say why the stone should elude our touch; it may be temporarily sacculated, in which case, to attempt its extraction would be very injudicious. Supposing, however, the stone is struck with the sound, the staff may then be re-introduced, and the operation proceeded with. Let me here call attention to a probable unpleasant effect of the introduction of the staff, viz. the emptying the contents of the bladder, by the side of the instrument; which renders the operation somewhat more difficult.

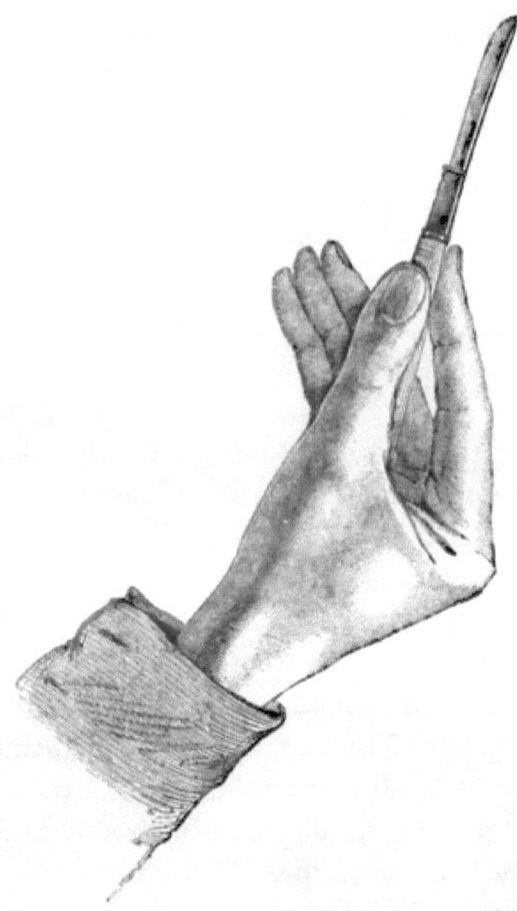

I generally prevent this, as far as possible, by squeezing the urethra against the staff; and being thus restrained in its action, the bladder soon ceases to contract. The surgeon, now seating himself in the low chair before mentioned, should feel the prominent points of bone at the outlet of the pelvis, and then receive the knife from the assistant. The knife I am in the habit of using, and the position in which I hold it, are depicted in the wood-cut on the previous page. It will be observed to be longer and more pointed than the common scalpel. The manner of holding it I believe to be very important; at least, I am sure it is, in reference to the comfort of the operator. The handle should be just long enough to rest in the palm of the hand, between the metacarpal bones of the index and middle fingers; and the knife should be, from the first, delicately but firmly held in that position.

By this means, any necessity for altering the mode of holding the knife, after the first incision, is avoided, and that incision, also, can more easily be made of the desirable bold and piercing character.

The surgeon now, having his left hand free, either to return a prolapsus ani, or hold up the scrotum, should plunge his knife into the perinæum, just below the scrotum, and about a quarter of an inch, or rather less, to the left of the mesian line. The point of the knife should be directed towards the groove of the staff, and may be steadily thrust directly backwards, if possible, into the urethra. If the urethra be pierced, a few drops of blood will issue from the penis. I believe a great part of the success of the operation depends upon the boldness and dexterity of this first step. The knife thus introduced, should then be carried downwards, in an oblique direction, to a point midway between the anus and the tuber ischii, the incision being made gradually more superficial. The index finger of the left hand should next be introduced into the bottom of the wound, to feel for the groove of the staff; and if the point of the knife have not pierced the urethra, an incision should be made into it, and the knife placed in the groove. Then, taking the staff from the assistant, the surgeon should depress the handle so as to make the groove serve as a director; and this depression, in the case of children, must be very great, as the bladder in them is situated so high in the pelvis. The knife should then be carried along the groove until the neck of the bladder is divided, which is known by the gush of urine. If the bladder be empty, the operator must rely upon his anatomical knowledge. Occasionally his knife may strike against the stone. When the staff is depressed, great care should be taken that its point is still in the bladder, and a slight onward motion may be given to it, at the same time that the knife is carried along its groove. A precaution is here necessary, that the knife should not be suffered to slip out of the groove, as I have seen it carried through the posterior wall of the bladder into the abdomen; with what results may easily be imagined.

As soon as the neck of the bladder is divided, the knife should be at once withdrawn, without any further division of the structures, and given to an assistant. The handle of the staff should then be taken in the right hand, while the index finger of the left is passed into the bladder, directed by the staff.

The incision being small, the finger will dilate it. The surgeon should remember, with reference to this step of the operation, that the prostate in children is very small and ill-defined. Hence arises, in part, the necessity for using dilation instead of extending the incision. If the neck of the bladder have not been divided (and this has sometimes happened), the greatest precaution will be necessary on the part of the surgeon to avoid passing the finger into the space around the bladder, for here the cellular tissue being loose, and yielding easily to the pressure of the finger, may give the impression that the bladder has been reached. The results of this error are, of course, most disastrous. To avoid it, it is necessary to feel distinctly, that the finger is within the neck of the bladder, and if possible, it is very desirable to touch the stone before withdrawing the staff. Should the staff have been withdrawn before the finger is in the bladder, it is useless to attempt to re-introduce it. The surgeon should be bold enough to abandon the operation for the time.

When the surgeon has ascertained that the bladder is opened, he withdraws the staff, and keeping his left fore-finger still within the bladder, he takes the forceps, well oiled and warmed, in his right hand, and passes them into the bladder, guided by his left fore-finger, which he at the same time withdraws. A sudden gush of urine impels the stone towards the opening, and if the forceps be quickly opened, it will sometimes fall into them. If this do not happen, the surgeon should carefully strike the stone with the forceps, and only when he has succeeded in doing so, should he open the blades and grasp it. There may be considerable difficulty in this part of the operation, particularly if the bladder be empty. The stone may seem to be lodged behind the pubes, in which case, pressure above the pubes, by the hand of the assistant, will frequently push it into the blades of the forceps. When seized, the surgeon should gradually and carefully withdraw forceps and calculus together. Should the stone be caught in its long axis, it should be dropped and again seized. Occasionally it is so small as to elude the grasp of the forceps, in which case the scoop will often serve the desired purpose. When the stone is withdrawn, the surgeon should re-introduce his finger to feel if there be a second; and the stone should be examined to see if any fragments have been broken off. If this be the case, they should be carefully removed.

Then the assistants should rapidly untie the patient, and a sponge dipped in cold water being applied to the perinæum, he should be placed in bed. A subject for anxiety presents itself at this stage, which I think has been rather exaggerated. It is the prospect of hæmorrhage. At the time of the operation a certain quantity of blood must necessarily be lost; nor is it of any use either to tie the small arteries which may be seen bleeding, or to search in the deeper parts of the wound for larger vessels. The bleeding ceases of itself; or if it continue, the child faints, and that puts a stop to it. In cases in which a trickling of blood persists, I keep the patient in a somewhat raised posture, that syncope may the sooner supervene. It is useless, and needless also, to apply pressure—sitting with one finger in the wound, as I have often done when a dresser — or to plug, or apply ice. There is scarcely ever any after bleeding unless the child is near the age of puberty; then, however, it may be very serious; and I have seen a case in which it proved fatal. In this instance the bleeding recurred several times, and there was probably a hæmorrhagic diathesis. The treatment of patients of this age is not within the scope of my design; but I should certainly, if possible, avoid lithotomy at that period of life. If the genital organs have passed out of the undeveloped state which characterizes them in childhood, I think it would be better to defer the operation until their evolution is complete.

It is not necessary to use any catheter or other contrivance for the withdrawal of the urine. This should be allowed to escape through the wound, and it is very advantageous to give these poor little fellows a good dose of opium, to quiet them while the urine is passing over the freshly-cut surface. The extreme sensitiveness and consequent excessive pain diminish in a few hours. If the flow of urine be impeded by coagula forming within the wound, these may be cautiously removed by the finger. The after treatment consists primarily in great attention to cleanliness, and especially to the constant removal of the draw-sheet, so as to keep the patient as dry as possible.

If the case progress favourably, the urine will begin to escape by the natural passages upon the eighth or ninth day. The wound gradually closes, and at the end of a fortnight the boy will be able to sit up, and before the end of the third week, if he have been previously in good health, he will begin to run about. Great care is necessary that these patients should not be over stimulated, as a tardy convalescence is the inevitable result. The diet should be milk and eggs, and fish, very little meat, and no beer. The patient, also, must on no account be permitted to run about before the wound is perfectly closed; as a fistula, extremely troublesome to heal, may result, and last for months.

I have once or twice had these little patients come back to me, after having left the hospital two or three months, on account of incontinence of urine; and upon inquiry, have found that they have been badly fed, their parents having been unable to obtain for them the necessaries of life. The treatment is obvious. A fortnight or even a week in the Infirmary has always perfectly relieved them.

To some of the *accidents* which are liable to occur during the operation of lithotomy, I have already referred; the possibility, namely, of the staff being forced through the walls of the urethra; of failure to open the neck of the bladder, and consequent introduction of the finger into the surrounding cellular tissue; of the knife, after opening the bladder, being carried through its posterior wall into the peritonæal cavity, or even into the iliac artery; and of leaving fragments of the stone, or a second stone, in the bladder. Besides these, there are others. The knife may slip out of the groove of the staff; or the rectum may be cut (generally in the first incision); or the incision into the bladder may be made too large, reaching the recto-vesical pouch of the peritonæum; or one blade only of the forceps may be introduced into the bladder, the other being pushed into the cellular tissue external to it. Of these, only the wounding of the rectum demands special notice; it occurs, as I have said, in making the first incision, especially if the bowel have not been well emptied of its contents. The accident may or may not be discovered at the time of the incision; most frequently it is betrayed only by the appearance of fæces in the wound. The treatment is to divide the sphincter ani, which should be done without delay.

Division of the artery of the bulb has been enumerated among the accidents to which the operation of lithotomy is liable. For my own part, I believe the artery of the bulb is always wounded, and I scarcely see how this can be avoided; nor do I know of any particular danger attending it.

If the stone be of very large size, laceration of the tissues may be unavoidable if it be extracted, and the consequences of this are very disastrous. Sloughing follows, with possible infiltration, and all its train of evils. In such cases, what should be done? I think the surgeon should provide himself with a lithotrite in every case in which the stone is large, so as to be prepared to crush the calculus, if necessary, previous to its removal. In this case a stream of warm water should be thrown into the bladder, to insure the removal of the fragments.

Sloughing of the Wound.—Now and then, a few days after the operation, or even as much as a fortnight or three weeks, without any assignable cause, the surface of the wound takes on a very unhealthy action; in fact, begins to slough. And this goes on with great rapidity. The constitutional disturbance is slight, sometimes there is even none. I have generally attributed this condition to over-feeding. At all events, I have several times relieved it by taking off the wine and meat these little patients were receiving, and substituting milk; and I now universally adopt the diet I have recommended above.

Occasionally there occurs after lithotomy a swelling of one of the testicles. This arises probably from the irritation of the urethra, and subsides in the course of a few days. I have also had one patient in whom condylomata appeared between the buttocks, doubtless as the result of the constant trickling of the urine. They disappeared directly the urine flowed through the natural channel.

Causes of Death after the Operation.—These are hæmorrhage, peritonitis, infiltration of urine, pyæmia, and visceral disease. I have already referred to hæmorrhage, which, indeed, is rare in children. *Peritonitis*, also, very seldom occurs, unless there has been some undue extension of the incision in the operation. If it should arise, opium is the great remedy; leeches, also, may be applied over the region of the bladder, and subsequent fomentation should be freely employed. The prognosis is unfavourable, but in respect to the diagnosis, it should be remembered that pain over the region of the bladder may arise from the presence of a clot obstructing the wound; but this occurs generally very soon after the operation, before peritonitis is likely to have supervened.

Infiltration of urine, or pelvic cellulitis, may originate from some undue manipulation, either by finger or forceps, during the operation: possibly it may arise without. In any case, it is a very fatal complication. A day or two after the operation, if severe symptoms of prostration appear, particularly if accompanied by restlessness, a quick and feeble pulse, and a glazy or sloughy appearance of the wound, with fætor, we may have serious suspicion that the urine is becoming infiltrated into the areolar tissue, and that death is almost certain. Can anything be done? We can only trust to supporting the general powers, carefully syringing out the wound to free it from the fetid sanious discharge. I can scarcely see any good result likely to ensue from laying the wound and the rectum into one, as has been recommended.

Visceral Disease.—In reference to this subject it is only necessary to mention particularly disease of the kidney. And even this is rare in the case of children under puberty. In consequence of the difficulty of diagnosis, it may be found to have been existing undiscovered, perhaps unsuspected, before the operation. In this case its symptoms soon make their appearance, generally under the form of anasarca. The issue is fatal. On post-mortem examination the kidney will be found to contain pus, either distending the pelvis, or diffused in points through the substance of the organ.

The following account of a fatal case of lithotomy in a boy seems to combine in one the various causes of death alluded to:— J. H., aged ten years, was admitted into Guy's Hospital on the 5th of February, 1859. He came from Lincolnshire, and had suffered from symptoms of stone all his life. For some years he had complete incontinence of urine. Three days after admission the operation of lithotomy was performed, and a large lithic acid calculus was removed. The boy never appeared to rally, and died at the end of six days. On the post-mortem examination, there was found general acute inflammation of the whole of the peritonæum. The liver was adherent to the stomach. The greater quantity of lymph was found in the lower part of the abdomen, where the intestines were adherent to the bladder. The left ureter was much dilated, as well as the pelvis of the corresponding kidney, the cortical structure of which was proportionally wasted. The right kidney was healthy. The bladder was very much diseased, and appeared contracted to the size of the stone which had been removed. Its mucous membrane was sloughy, as was also the external wound, and the cellular tissue around, which were infiltrated with serum and purulent matter.

CHAPTER XI.

THE EXTERNAL GENERATIVE ORGANS.

DISEASES OF THE TESTIS.

THE affections of the testis are divisible into those of the tunica vaginalis and cord, and those of the testicle proper.

I. DISEASES OF THE TUNICA VAGINALIS AND CORD.

1. *Hydrocele of the Tunica Vaginalis.*—This is very common in children, and may be met with at any age, from a month after birth and upwards. The mother brings the child full of alarm lest it is ruptured, and the diagnosis is not always very easy, particularly if the hydrocele is of the kind known as congenital. The form of the scrotum is globular, and not, as in adults, pyramidal. The quantity of fluid contained within the tunic varies from two drachms to an ounce and half; it is almost colourless, and permits light to pass through it with the greatest ease; the skin also is extremely thin. Nothing, therefore, can be more perfect than the transparency of hydrocele in a child.

The disease consists, as is well known, in the

effusion of fluid into the tunica vaginalis. It presents varieties dependent upon the closure, or non-closure, of the vaginal process of the peritonæum. In the latter case it is called congenital, and is to be distinguished from the common form by the fact, that the contents may be gradually squeezed into the abdomen. I have never met with a case, in children, of what is called encysted hydrocele. In respect to the diagnosis, this affection can only be confounded with hernia. The common variety is easily distinguished, it is transparent; it is not returnable into the abdomen; and the swelling can be entirely isolated by squeezing the upper part of the scrotum between the fingers. The congenital hydrocele resembles hernia in being returnable into the abdomen; but it receives scarcely any impulse on coughing; it is transparent, and after it has been returned, if the patient be made to stand erect, and the ring be closed by the finger, the fluid will gradually re-accumulate in the scrotum. The fluid also can only be pressed into the abdomen slowly, not returning at once like the bowel: in some cases, it requires patience to return it at all.

The *Treatment* of hydrocele in children is very simple. The first plan that should be adopted in every case is the application, to the scrotum, of lint dipped in a lotion of hydrochlorate of ammonia (an ounce to a pint and a half of water), or even in cold water; the child being kept in bed. Very frequently, with nothing more than this, the swelling disappears entirely. If, in a week's time, no diminution ensue, a puncture may be made with the point of a lancet, or the fluid may be drawn off by a small trocar and canula. Or a grooved needle may be passed into the tunica vaginalis, and its interior scratched with the point of the instrument. Whichever of these plans is adopted, I have, as often as not, found the fluid in as large a quantity as ever on the following day. These methods failing, a different mode of treatment must be resorted to in the common and congenital hydrocele respectively. In the common form, in which the vaginal process is closed, I pass one or two threads through the tunica vaginalis, and tie them, to form a seton. They must be removed on the second day. In performing this simple operation, it is necessary to take care that the tunic is made quite tense, and that the needle carrying the thread does not, as I have seen it, pass between the tunic and the skin. Inflammation should follow; it is attended with considerable swelling, and sometimes a good deal of constitutional disturbance. The swelling subsides in the course of a few weeks, and the patient remains well.

In congenital hydrocele, pressure at the abdominal ring, by means of a truss, is the course usually recommended. When, however, the difficulty of passing the fluid into the abdomen has been great, and consequently the communication may be believed to be very small, I have used the seton. In one such case slight peritonitis supervened, but the seton was kept in for four days, which I now think is too long a time. The patient rapidly got well on its removal.

2. *Hydrocele of the Cord.*—This is an affection resulting, as I believe, from a partial non-obliteration of the vaginal process of the peritonæum, and is comparatively rare. It presents itself as a small swelling in the inguinal region, resembling a bubo, or a hernia in an early stage, and consists in an accumulation of fluid within a space which is isolated both from the abdomen above, and from the general cavity of the tunica vaginalis below. Being fluid, its contents give a feeling of fluctuation; and in some cases the swelling may be seen to be transparent. The diagnosis from hernia is somewhat difficult; the fluctuation, and incapability of being returned, are the diagnostic symptoms. But when the hydrocele is small, it may often be made to disappear under the abdominal walls, simulating the return of a hernia.

The best treatment is to leave it entirely alone; unless it should become very troublesome, when a needle may be passed into the swelling, and its contents evacuated. These re-accumulate, but slowly. Hydrocele of the cord generally causes very little inconvenience, unless it increases in size. In this case, obliteration of the sac should be attempted, not by injections, but by the use of a seton.

Case.—G. W., aged three years, was sent to me in May, 1857. Twelve months before, the mother first noticed a small swelling in the left groin. This had gradually increased in size. A truss had been applied. I found an elastic moveable swelling in the left inguinal region, which could be apparently slipped up and down along the cord, no doubt from the skin slipping over it. It was transparent, and of course protruded below the truss. There was no fulness at the internal ring. I tapped the swelling, and drew off more than two drachms of fluid; it quickly filled again, and I repeated the tapping. The third time I passed a seton of two threads through the hydrocele, and withdrew them at the end of the second day; at the expiration of a month the child was quite well. Friction with the hand over these swellings has now and then seemed to cause absorption, and an ointment of iodide of potassium has been frequently used successfully by my colleague, Mr. Poland, who has reported several cases in the *Guy's Hospital Reports* for the year 1851.

3. *Hæmatocele.*—A collection of blood in the tunica vaginalis, in children, may arise from a blow, or from the puncture of a vessel. As in adults, the swelling forms slowly, not from the continued effusion of blood, but from the gradual thickening of the tunic around the effused mass. This thickening is more distinctly to be observed in the child than in the adult, and it may proceed to the extent of nearly half an inch. If not attended to at once, this disease may continue for years; the swelling growing as the patient grows, until at last, from the obscure history, size, and weight of the tumour, it may be mistaken for malignant disease, and castration may be performed.

The *Diagnosis* of this affection is confessedly so difficult, that unless the history be most perfect, it is almost impossible to distinguish between it and certain diseases of the testicle. Hence arises the rule of treatment—the same for children and adults—always to cut into any swelling of the testicle before removing that organ; and if it be a hæmatocele, to turn out the contents, fill the cavity with lint, and allow it to suppurate. A hæmatocele is known, when cut into, by presenting, instead of the indications of malignant disease, a central mass of partially decolorized coagula, surrounded by the thickened tunica vaginalis.

II. DISEASES OF THE TESTICLE PROPER.

1. *Orchitis, or Swelled Testicle,* generally arising, as it does in the adult, from gonorrhœa, is of course rare in children. It is therefore not my intention to enter at large upon the subject. As I have observed, in speaking of lithotomy, it may occur as a consequence of that operation. It may also occur as a metastatic affection, in connexion with cynanche tonsillaris.

Case.—F. F., aged twelve years, was brought to me on the 14th June, 1850, by his father, who was full of suspicion as to the cause of a swelling of the left testicle. The swelling had commenced five days before, he having previously to that had cynanche tonsillaris, which had suddenly subsided. The epididymis was acutely inflamed, very tender, and hard; and the scrotum was red and swollen. There had been no discharge from the urethra. Knowing the effects of quinine in cynanche tonsillaris, I gave the boy two grains three times a day; applying also four leeches to the testicle. He was brought to me on the 21st of June quite well. There was no discharge from the urethra, nor any further affection of the throat.

2. *Scrofulous and Malignant Disease of the Testicle.*—Until the testicle has been roused to the performance of its peculiar function, structural disease of it, in the form either of scrofulous or malignant enlargement, is comparatively rare. The diagnosis between the early stages of these two forms of disease is almost impossible. In both, the swelling is irregular, soft, and semi-fluctuating; and until the diseased mass discharges itself externally, there is no certain sign by which they may be distinguished. Both of them, however, in time, soften and ulcerate; and then, the diagnosis is generally easy. The scrofulous swelling discharges a yellowish cheesy matter; the malignant, fungates and bleeds.

Case.—Strumous Disease of Testicle: Operation: Cure.—H. B., aged six years, was admitted into Guy's Hospital, January 6th, 1858, under Mr. Birkett's care. He was a very delicate boy. Two years ago he had an abscess over the left trochanter, which healed after six months. Then the left testicle began to swell, and became painful. The scrotum soon inflamed, and, six months before admission, gave way. The opening had been very slowly, but gradually enlarging ever since. On admission, the left testicle was exposed; it was about the size of a walnut, and covered with granulations. The margins of the wound were irregular, excavated, and in some places thickened. There was no pain, except when the part was disturbed. No enlargement of the inguinal or other glands. The right testicle was healthy. At the end of a month, though the boy's health was improved, there was no change whatever in the condition of the testicle and surrounding parts, and the disease was spreading. The mass was accordingly removed. The wound was a considerable time in healing, but the boy went out at last quite well. On examination of the testicle, it proved to be a well-marked case of strumous disease.

The difficulty in diagnosing this disease was great at the commencement: so much so, that no other course of treatment was applicable than the one that was adopted. The protruding mass was very unlike the structure usually presented by the scrofulous testicle when the skin has given way.

When malignant disease of the testicle exists, should it be removed? Certainly not with any view of saving the poor child's life. Such cases are sure to terminate fatally in a few months, from development of the disease in other parts. But excision should, nevertheless, be performed. Under chloroform it is painless; and it is a much preferable death to succumb to internal than to external cancer.

The Operation of Castration.—An incision should be made along the whole length of the swelling, from over the external ring to the bottom of the scrotum. The cord should next be detached from the surrounding parts, and a strong ligature *very* tightly tied around it. As soon as this is done, the cord should be divided about half an inch below the ligature, and the testicle dissected out. Any bleeding vessels (which can only be in the scrotum) may be tied, and a piece of lint, dipped in cold water, applied over the whole.

It will be observed, that this is a modification of the usual operation; and, as it must be allowed, it is a much simpler plan. There is no risk of the temporary ligature slipping off the cord; nor have we any concern lest the latter should retract, and compel us to search for the bleeding vessels within the inguinal canal. The only question is, whether any ill results of pain or irritation arise from including the whole cord within the ligature. As yet, I have seen none. The application of a ligature, with the greatest possible tightness, to any nerve, seems to destroy its sensibility.

III. MALFORMATIONS OF THE PENIS.

Hypospadias, Epispadias, Congenital Phymosis.—When the under wall of the urethra is deficient, the condition is known as hypospadias; when the upper wall is deficient, it is called epispadias. The former is, by far, the more frequent affection, and the defect may exist in very various degrees, from a slight incompleteness at the orifice, to total absence of the inferior wall of the penis. When epispadias exists, there is generally also a deficiency of the lower part of the abdominal wall; a most miserable condition. The posterior wall of the bladder then appears on a level with the anterior abdominal parietes; the urine, instead of being retained in that viscus, flows directly over the pubes, which it excoriates.

In neither of these conditions, can efficient relief be given by surgery. An autoplastic operation is of no benefit. In epispadias, a mechanical contrivance may be tried, in the shape of a piece of gutta percha moulded to the lower part of the abdomen; but this answers very imperfectly. However accurately it is made to fit, the water soon escapes by the side.

Whilst on this subject, I may mention a very curious case of congenital malformation which I saw in a child, aged eight months, who was sent to me by my friend Dr. Willshire, in February, 1857. The penis was (without erection) in a state resembling chordee, bent like a syphon. The only explanation that could be obtained from the mother, was that whilst pregnant, her husband had gonorrhœa, and suffered much from chordee, which gave her great disturbance and anxiety. A small portion of the prepuce was adherent to the glans; this I separated, but without any effect on the bent condition of the penis, which evidently resulted from some change in the body of the organ.

Congenital Phymosis is a constriction of the orifice of the prepuce, which prevents its being retracted over the glans. Sometimes the constriction exists to such an extent, that the orifice will scarcely admit a small probe; and then, every time the urine is passed, it swells out the prepuce into a bag. In this case, as I have previously observed, the retention of urine within the prepuce may give rise to the formation of a calculus at that spot; or to symptoms simulating those of stone in the bladder, such as frequent desire to micturate, and difficulty in the act; also, the little patient will frequently seize the prepuce, as if it itched, and elongate it. It may be noticed, however, that the urine is quite natural, and that it does not deposit any mucus. When the penis is examined, it is found that the attempt to draw the prepuce over the glans is attended with great pain, so that even the orifice of the urethra can scarcely be seen. In the cases in which the whole of the prepuce is distended with urine, each time the child passes water, it frequently passes over some ulceration of its inner surface, which may account for the pain in micturating.

The treatment consists not in simply slitting up the prepuce, but in circumcision. This is best performed by seizing the lengthened prepuce with a pair of dressing forceps, applied transversely, just beyond the extremity of the glans, and cutting off the redundant skin with a pair of sharp scissors; chloroform, of course, having been administered. Then, drawing the skin of the penis as far back as possible, the internal layer of the prepuce will be seen covering the glans. This layer must be slit up on a director as far as the corona glandis, and turned back so as to expose the glans. Frequently, this layer of the prepuce is found, more or less, or even entirely, adherent to the glans; and in this case, it must be separated from it in its whole extent. This should be slowly and efficiently done, and the groove behind the corona completely exposed. It may also be advisable, if the frænum be tight, to divide it at the same time. Care should be taken in these cases, that the director does not slip into the urethra instead of between the prepuce and the glans, as I once saw it do, and the urethra slit up in consequence. Formerly it was my practice to unite the internal layer and skin of the prepuce by sutures, but latterly, except in a few cases, where the distance between them is very great, I have discontinued this practice. If the edges are brought together as closely as possible, and wet lint applied, union will take place by granulation. Should sutures be employed, the uninterrupted suture, which I first saw practised by Mr. Cock, will be found the best.

My reason for performing circumcision in these cases is not only to relieve the present symptoms, but also because this operation, by exposing and hardening the glans, and facilitating cleanliness, diminishes the risk of disease in after life. Three weeks or a month generally elapse before the parts are entirely healed. Sometimes slight erysipelas may follow, or an abscess form in the skin of the penis; and occasionally, if a sufficient portion of the prepuce have not been removed, the skin that has been drawn back over the glans, contracts and forms a constriction, constituting a species of paraphymosis, the next subject to which I shall allude.

Paraphymosis.—This affection is a constriction of the prepuce behind the corona glandis; not congenital, and indeed almost always, in children, accidental.

A boy pulls the skin of his penis back (or one may do it for another), and finds a difficulty in returning it; then, being ashamed or afraid to speak, he suffers it to remain, and swelling, both of prepuce and glans, at once begins. The longer the delay, the more this increases, until at last, unless relieved, the prepuce begins to ulcerate. The ulceration will continue until it has relieved the tension; the swelling then subsides, leaving an extensive raw surface behind the glans, which occupies some weeks in healing. As cicatrization takes place from the consequent contraction, the penis becomes in as bad a state as before, and the process recommences. How far this might be carried, if left alone, I cannot tell; it is said that the urethra may be ulcerated through, and urinary fistula produced, but this has not come within my experience.

However this affection be produced, the treatment is the same, viz., to draw the prepuce forwards. And the best mode is to place the boy on his back, to grasp the penis firmly with the right hand, and proceed to lift him bodily up. The prepuce invariably slips forward, and the affection is at once relieved. There is no danger in the process; the penis can bear all the force that could be so applied. I now never have recourse to any other plan; division by a director is unnecessary; pulling the prepuce forward while the glans is pushed back, tedious and inefficient. When the prepuce is thus restored to its position, a state of phymosis is produced, which, under the use of cold applications, subsides in a few days.

IV. GONORRHŒA.

Strange as it may seem, I cannot, in treating of the urinary organs of male children, entirely pass by this subject. I have seen three cases of well-marked gonorrhœa in little boys between four and seven years of age. In each case the history given, and the mode in which the affection was detected, indicated that it was acquired in the usual way. And it is necessary to be aware that the position of nurse is sometimes, though happily very rarely, held by persons through whom the contagion may be spread. In one case I had to cut off a crop of warts in a boy, and in that instance the same operation was necessary for the person under whose care he had been placed.

The treatment is similar to that required for adults; but happily the cure is generally easy. Mild astringents suffice; the best I have found is the following:—

℞ Glycerine . . . . . . ℨj.
Liq. Plumbi . . . . . ℨss.
Aquæ . . . . Oss.

To be used as an injection, two or three times a-day, after the inflammatory symptoms have subsided.

V. STRUMOUS ULCERATION OF THE PENIS.

I have never seen a case of primary syphilis in boys before puberty, though doubtless it might occur. There is a condition, however, demanding notice, of which the following case is a very good illustration:—

Case.—C. S., aged three years, was admitted into the

Royal Infirmary for Children, in April of the present year. A small strumous, ill-conditioned child. His mother states he has been ill since birth, having early taken the usual infantile diseases. There exists an irregular-shaped sore on the glans and prepuce, on its under surface, around the frænum. The sore is very superficial, and has existed to a greater or less extent for the last two years. It has never quite healed. Three of his brothers are or have been so troubled, up to three or four years of age. The girls of the family have all been free from any strumous ulceration of the labia. By good diet, and constant application of nitrate of silver to the sore, which was merely to prevent irritation from the urine, the child got quite well in a month.

VI. ONANISM.

This unfortunate habit often commences, both in boys and girls, at a very early age; certainly between three and four years, if not earlier. At this age it must be regarded as a purely physical disease; but I do not know of a remedy. Of course, all hygienic means should be adopted; early rising, plenty of exercise in the open air, and special cleanliness in respect to the organs concerned must be insisted upon. The least accumulation of secretion seems to operate as a strong excitant. Beyond these general means, and the serious but kind exhortations of the parent, I know of nothing that can be wisely done. All irritating applications, which produce soreness of the parts, lead to the aggravation of the disease. The hope must be that the inclination may pass off under this general care; or that, at the period when moral suasion can be employed, which comes sooner than might be thought, it may be found successful. The effects of the habit upon the health are injurious, but not necessarily irremediable. Surgeons should bear it in mind whenever children fall off in their appetite, and look pale or languid, without obvious cause. Probably the same physical conditions which induce this practice, may lead to dreams upon the part of children, which dreams, as before suggested, may have their share in the extraordinary and almost unaccountable charges sometimes made by them against their friends or guardians.

CHAPTER XII.

HERNIA.

I. UMBILICAL HERNIA.

HERNIA, at one or other of its usual seats, is a common affection in children. It is most frequently seen at the umbilicus; next in the inguinal region; rarely in the femoral. When we remember that the umbilicus is the part of the abdominal wall last closed, and that till the very termination of fœtal life it presents an opening for the passage of vessels, we cannot wonder that the umbilical is the most frequent form of hernia, nor that the protrusion should sometimes contain an enormous mass of intestine. This may amount, I believe, to a full half of the abdominal contents.

Case.—A boy, aged eighteen days, born of syphilitic parents, was brought to me at Guy's Hospital in 1856. The abdomen presented a swelling of immense relative size, certainly not less than that of the patient's head. The mass was irregular in form, and a large part of its surface was covered with a green dry slough, as if the contents were about to protrude. The integuments of the remaining part were exceedingly thin. At the base of the swelling a distinct ring could be felt, formed by the parietes of the abdomen: this ring was over two inches in diameter, and had a very defined margin. I took the little patient (with its mother) into the hospital, with the full expectation that it would soon be dead. The contents of the swelling were as far as possible returned; a bandage was applied round the abdomen, and good nourishment given to both mother and child. To my surprise, the slough separated, the swelling became smaller, healthy granulations arose from the surface of the sore, the child improved in health, and at the end of six weeks the whole surface had cicatrized. The little patient went out with the swelling about half its former size, though still, of course, requiring support. I have seen this child several times since, and have observed the swelling gradually to subside, I believe partly from the contraction of the cicatrix. The ring also has been smaller at each time of observation. There is now scarcely any protrusion, and the ring only admits the tip of the little finger. Unfortunately the child has become the victim of disease of the dorsal vertebræ.

By far the most frequent form in which this complaint is presented to our notice, is that of a rounded or oblong projection from the umbilicus, which gives a peculiar pulpy feeling to the fingers; evidently that of a portion of intestine containing air. The contents of the swelling are easily returnable, and when returned a distinct ring, large enough to admit the fore finger, can be felt in the abdominal walls. When the finger is withdrawn, the swelling immediately protrudes again, and, should the child happen to cough, an impulse is received. Perhaps one reason why the intestines are so easily felt in these cases is the absence of the fascia propria, which, lining the general abdominal walls, is here deficient; hence there is nothing in front of the intestine except the skin, the superficial fascia, and the peritoneum. Whenever the child cries there is, of course, an inclination to a still greater protrusion. One of the great principles of treatment, therefore, consists in supporting this weakened spot, especially during the act of crying:— not, as I think has been very unwisely recommended by putting a nipple-like body into the opening, which must tend to retard its closure, but by applying a flat pad of lint with a broad band of strapping around the abdomen to fix it, and a roller over all. Modifications of this plan may suggest themselves; a piece of lead instead of lint, and various kinds of elastic bandages, have been recommended. If these applications are kept constantly to the parts, the swelling diminishes, and in time the opening completely closes. But I have also seen cases in which the cure has been quite as perfect when no care whatever has been taken. I have never known an operation required for strangulation in this form of hernia. The protrusion in these cases does not take place exactly in the position occupied by the umbilical vessels, but a little above it.

II. INGUINAL HERNIA.

In children, inguinal hernia is usually congenital; that is, the tunica vaginalis has not closed, and the bowel descends in this process of peritoneum. The hernia accordingly has no proper peritoneal sac, and if it pass into the scrotum, lies in contact with the testicle; or, if the testicle have not completely descended, the hernia lies in contact with it in whatever part of the canal it happens to occupy. Sometimes the testicle forms a kind of plug in the inguinal canal, preventing the descent of the gut; as the gland descends, the hernia protrudes with it. A "hernia infantilis" is also described by authors, in which the bowel is said to pass *behind* the still open tunica vaginalis, carrying the peritoneum before it. This constitutes a hernia covered by three distinct peritoneal layers, two of which belong to the unobliterated vaginal process, while the other forms the proper hernial sac. In my experience, I have never met with a case of this kind in children; nor do I think it occurs until adult life, although it is unfortunately called "infantile."

The *Diagnosis* of hernia in children is a matter of some importance. The affections which need to be distinguished from it are, in the very young, an undescended testicle, and hydrocele; and, as the child advances towards puberty, enlarged glands in the groin, and psoas abscess. When these affections are uncomplicated, the distinctive marks are too evident to need further description than has already been given; but it may happen that hernia may co-exist with one or other of them. For example, there may be both a hernia, and a hydrocele of the tunica vaginalis, on the same side. In this case, the upper boundary of the hydrocele may be distinctly felt; and the hernia, only, is returnable; an impulse may also be felt on coughing. Or a hernia may co-exist with an undescended testicle. In such a case the peculiar sensation occasioned by squeezing that organ, and the peculiar feeling to the surgeon's touch, combined with the possibility of returning part of the swelling, indicate the nature of the case.

The hernia most frequently met with in children, in this region, is the oblique inguinal. This forms a swelling either in the groin or scrotum, having the characteristic and wholly peculiar consistency of protruded intestine; it disappears when the child lies down, and increases considerably on any exertion, particularly on crying. It can be retained in position by the finger placed in the inguinal canal, which, as I have before observed, cannot be done in the case of congenital hydrocele.

The *Treatment* of hernia in very young children is a matter of considerable difficulty. The end to be sought is the closure of the vaginal process of the peritoneum, which will generally take place if the intestine is kept within the abdomen. How best to accomplish this, is the first question to be considered. I believe that, until the child runs about, a pad and figure-of-8 bandage applied by the nurse, who should be carefully instructed to do it, is the best contrivance. I have seen trusses applied at this early age, and a great deal of excoriation result from them. When the child can run about, however, any mechanical contrivance that will keep the hernia "up," whether a steel spring, or air pad, or an elastic webbing, may be used. My own preference is for a slight steel spring; and, if it be well kept on, before the child outgrows it, the hernia will be cured.

If the hernia be complicated with hydrocele, the contents of the latter should be evacuated, and a truss applied in the same way. If, with the hernia, there be an undescended testicle, the case is better left untreated until the testicle is so far descended as to allow the hernia to be completely isolated, and a truss applied to it without pressure on the gland.

A question may arise respecting the operation for the "radical cure." I should say this procedure was quite unjustifiable in children. Under the careful use of the truss almost all cases cure themselves. It is time enough to adopt severer measures when the patient arrives at adult life.

Strangulated Hernia.—Unfortunately, even in the youngest children—Mr. Curling met with a case at six weeks old—the intestines may, now and then, become strangulated. The symptoms are similar to those in the adult; and care should be taken, in every case of repeated vomiting in children, that the abdominal rings are free. As in adults, if the hernia cannot be returned while the child is under the influence of chloroform, an operation for the relief of the strangulation should be at once resorted to. It is not my intention to describe the method of performing it, as it does not differ from the same operation in adults. The surgeon should, however, recollect how thin, and how distinctly defined, the various layers of tissue are in children. And it should also be borne in mind that the contents of the hernial sac are almost exclusively intestinal. Any one familiar with the appearance of the infantile abdominal viscera will recollect how short and transparent is the omentum.

As the child approaches puberty, hernia may occur in the same way as in adults; there may be either ordinary umbilical hernia; oblique or direct inguinal, ventral, or femoral hernia. These, however, do not need any special remark.

CHAPTER XIII.

NÆVUS.

NÆVUS maternus is the name applied to a peculiar form of disease well known to surgeons, and I shall, therefore, always make use of it, notwithstanding the numerous synonymes it has received at various times. John Hunter called it aneurism by anastomosis; it has also been called Hæmatoncus, Teleangeiectasy, erectile tumour, with numerous other cognomens, serving only to distract the student.

The type of this form of disease may be considered to be the common subcutaneous nævus.

These tumours vary considerably in size and shape, and the part of the body upon which they appear is equally various. They are more frequently met with on the head and neck, than on the trunk or extremities; though no part can be considered to be free from them. John Bell asserts that they appear upon the viscera, and in the museum of Guy's Hospital there is a preparation of a peculiar vascular growth developed on the jejunum, which bears a close resemblance to structure of the nævus. It is the opinion of some pathologists, that similar growths are not very unfrequent on the liver and other internal organs. The external nævi vary exceedingly in size. They may be found covering the whole of an extremity, or nearly one half of the trunk of the body. On the other hand, a small red spot, not larger than a pin's head, may appear shortly after birth, upon some conspicuous part of the body, and remain of the same size to the end of the patient's life. At other times nævi gradually increase as the patient advances in age. They present every possible variety of shape, and are supposed to represent all kinds of different objects, according to the longings of the parent of the child upon whom they may appear. For instance, they are likened to slices of bacon, port wine, leaves, spiders, all sorts of fruit, mulberries, strawberries, &c. An attempt at classification has even been made by those who wish to give a name to every form; and they have thus been designated, nævus araneus—foliaceus—morus—ribes. I do not think their connexion with some object impressing the mother's imagination during the period of gestation, is an idea entirely to be discarded, though I would not, certainly, be supposed to credit the extraordinary tales related by many women. There is no doubt, however, of the mind being in a peculiarly susceptible state during that period, and this, to a certain extent, accounts for the fancies then indulged in. As a further illustration of the likeness of these marks to different kinds of fruit, it is said that they always become red at the time when the ripening of the fruit takes place. This, I need scarcely say, is easily explained, by the fact of the summer being the time at which the fruit ripens, and also the time in which there is increased action in the skin, and consequently greater vascularity in the swelling.

What is a nævus? It is a structure resembling erectile tissue in all its characters, but is not endowed with the power of filling itself with blood. It cannot, therefore, be rightly termed an erectile tumour, as some have called it, although, in point of *structure*, strictly analogous. It usually commences as a small spot of a pink or blue colour, first observed by the nurse, and commonly about one month after birth. For the most part, it gradually increases as the child grows, and becomes more and more vascular; small vessels seeming to diverge from a central spot. Upon pressure, these vessels may be emptied entirely of their contents, but they quickly re-fill when the pressure is removed. These changes are distinctly visible to the eye, when the nævus is cutaneous; when it is subcutaneous, the condition of the individual vessels cannot be seen, but the swelling disappears on pressure, and returns on its removal.

In colour, the cutaneous nævi are of a vivid red, the sub-cutaneous dimly blue. Both varieties become purple when the patient is out of health. When situated on the face, they swell and become more vivid in colour on a forced respiration; a similar effect is produced by heat, or by any considerable exertion. They have been described as presenting a bruit to the ear and a thrill to the touch; but unless they approach very nearly to the large arterial tumour, to which the term aneurism by anastomosis is appropriate, these symptoms are not present. On the chin or neck, or in any position where they can hang down, nævi, if large, become somewhat pendulous. They are not painful; if the skin be involved, they occasionally bleed when the child cries.

At what time does a nævus commence? I have already said that it is generally first observed about a month after birth. It has been assumed that it must have existed prior to this time, and therefore it is usual to speak of nævi as congenital. But I do not know of any sufficient grounds for this assumption; indeed, I think it more probable that they originate about the period of their discovery. Nurses and mothers are very careful in the examination of infants, and it is highly improbable that even a spot should escape them. The usual rapid growth of these formations, also, after their detection, would lead us to suppose that they originated about that time. It may be worth consideration whether some of the minor accidents to which infants are liable, as a bruise, or a prick with a pin, &c., may have any share in calling them into existence. Nævi, however, are certainly sometimes congenital; a small red spot may be observed upon some part of a child immediately after birth. Curiously, these spots seldom grow as the ordinary nævi do, and the name of *spurious nævus* has been applied to them. Indeed, they not unfrequently disappear. A case is related by my friend Mr. Caleb Rose, in the *Medical Gazette* of March 20th,

1846, in which two of these marks, situated one on either side of the nose in an illegitimate child that died, gave rise to suspicions of violence, and led to an inquest. In another number of the same journal, a case is mentioned in which a mark, of a bright-red colour, was observed half round the neck of a newly-born child, as if a ligature had been placed there. In such a case as this, however, a possible pressure of the cord having taken place should not be forgotten.

Pathology.—I have already said that the true type of this affection is the ordinary subcutaneous nævus. It will be necessary, therefore, that I should describe the structure of these formations. We only now and then have the opportunity of examining them in their entirety; as for the most part they are got rid of by other means than excision. The lower figure in the accompanying lithograph represents one that had been removed, and "cleaned," as an anatomist would say. It is the very specimen on which Mr. Birkett's admirable description of the structure of nævus, in the 30th volume of the *Medico-Chirurgical Transactions*, was based. According to that gentleman, the following elements composed the growth—viz.,

    1st. Areolar uniting or fibrous tissue.

    2nd. Epithelium.

    3rd. Capillary vessels and vessels of larger calibre.

The mass appeared to consist of lobes. Each lobe possessed a distinct proper capsule, which at its neck became intimately blended with the fibrous tissue of the true skin. Strong but delicate fibrous bands intersected each lobe in every direction. In some of

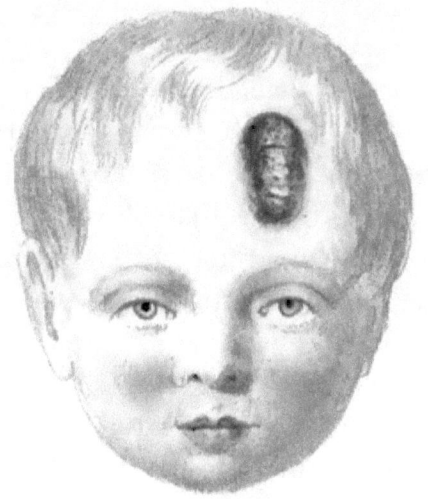
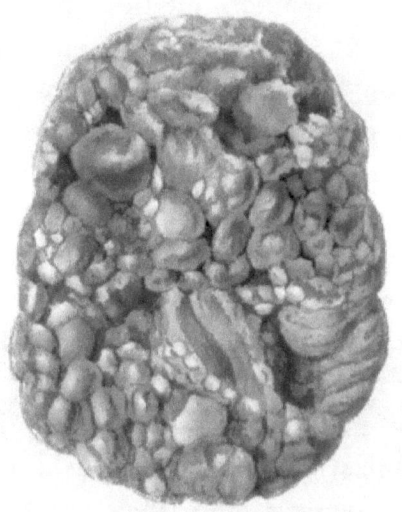

the lobes these bands seemed to have a characteristic arrangement; a dense and well-defined central point existed from which the bands passed off in a radiating manner to the investing sheath. The interior of the lobes presented, on a transverse section, a reticulated character. The cells composing them communicated with each other on all sides, so that air would freely permeate the whole lobe; but the lobes did not all appear to communicate with each other. The various septa, or bands separating the lobes, were composed of delicate fibrous tissue, covered with epithelium.

In respect to the vascular arrangement, by far the most important, the structure more nearly resembled that of the corpus spongiosum than any other tissue of the body. Each lobe was supplied by one or more distinct arteries, which, dividing into capillaries, ramified and were lost upon the septa. How the veins commence has not been clearly made out. Upon injecting these structures, however, they appear like one mass of blood-vessels; and if they are cut into, blood seems to flow from every point. My own opinion is, that the cells themselves are dilated veins.

Nævi appear to be the reverse of ordinary tumours; the vascularity in their case preceding instead of following organization. They seem, in fact, to be overgrown blood-vessels, and they result evidently from a defect, and not from an excess, of substance. The circulation in them is slow, a necessary result of the large calibre of the vessels; probably from this cause their vitality is low, and they readily ulcerate or slough in debilitated conditions of the system. If this take place, hæmorrhage, even to a fatal extent, may ensue.

In process of time, nævi, if not destroyed, may undergo a spontaneous change, called cystic degeneration. In this case the blood-vessels seem to become atrophied, the interior of the cells being filled with a clear fluid. The structure shrinks somewhat and loses its dark colour, the skin covering it presents a warty appearance and frequently exudes a watery fluid. In this condition it does not bleed when cut into.

Mr. Paget, in his admirable work on Surgical Pathology, classifies nævi into the arterial, the venous, and the capillary varieties. But, for practical purposes I am in the habit of dividing them into three classes, according to the structures invaded by the disease; viz. the subcutaneous, the cutaneous, and the mixed nævus. The last-named form is by far the most frequent, and consists of a combination of the other two.

Before proceeding to the treatment of these cases, however, I must say a few words on the *diagnosis* of the subcutaneous variety, which is not always perfectly easy, particularly if the nævus be covered, as it frequently is, by a mass of condensed areolar tissue. When situated in the neck, a nævus may be mistaken for one of the cysts which are occasionally developed in that locality, and known as hydrocele of the neck. The distinction, indeed, is often not less difficult than important; and great caution should always be used in dealing by operation with cysts in the neck, lest by

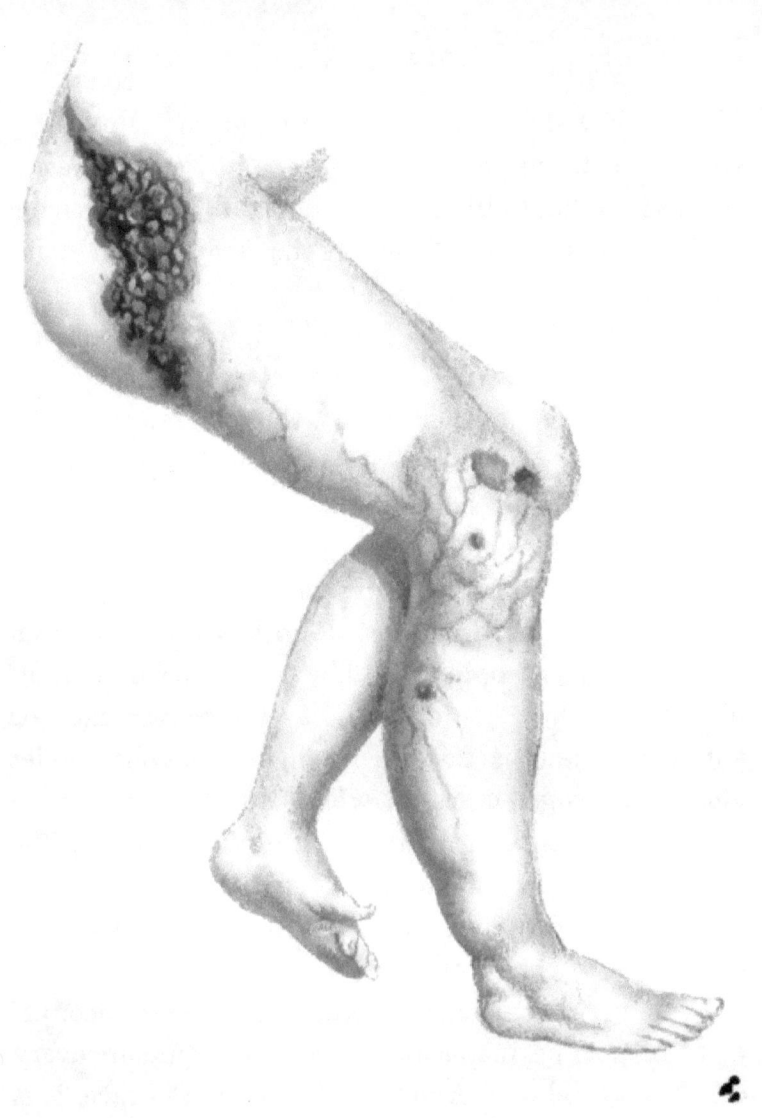

chance they should turn out to be nævi. Medullary cancer, also, might be mistaken for subcutaneous nævus; its fluctuating character, and our appearing to be able to almost empty the swelling in some cases, might lead to the error.

The Treatment of Nævus.—The first subject for consideration is, whether any treatment should be adopted at all: in other words, the general principles which should guide us in determining to operate. It would be folly to suppose that, because an anxious mother brings a child with a mark upon some part of its body, therefore the surgeon must of necessity perform an operation. Yet doubtless such has been a very prevalent opinion. The obvious rule is that, if a nævus does not occasion any inconvenience or annoyance, and is not increasing, it should be let alone. In some cases, in which a nævus occurs in the face, but is not growing, it is advisable to ask the friends whether they would prefer the existing red stain, or the white scar of an operation. My advice to them is to choose the first.

If, however, a nævus, whether at present inconvenient or not, be increasing in size, it should in every case be got rid of. And if it be upon the face, it is very important that this should be done at the earliest possible period. Only one circumstance limits the application of this rule, and that is, a very large extent of surface being involved in the disease. Nævi will sometimes extend over a great portion of a limb, as shown in the accompanying lithograph, in which the disease involved nearly all the lower extremity.

In such cases as these, of course, operative proceedings are out of the question. Nor, indeed, are they necessary; for in these extensive cases degeneration generally takes place, and the disease subsides into a harmless condition. In the foregoing lithograph the commencement of this process is seen, the upper portion of the nævus having become converted into cysts oozing a glairy fluid. This drawing was taken from a patient under the care of Mr. Bryant.

When the nævus is situated on the face, however extensive it might be, if it were spreading, I should be inclined to try and arrest it by repeated operations. Were it placed near the angle of the eye, especially, I should act on this principle, since the eyelids are very apt to become involved, and in them the disease spreads with extreme rapidity. In the lower of the accompanying lithographs, taken from a child under Mr. Birkett's care, is shown a nævus which spread most towards the lower lid. It might be asked whether, in the case of very large nævi, rapidly extending, their extension might be wholly or partially prevented by operations confined to their external portions. For my own part, I think no good result would attend this plan, and that a nævus must be wholly destroyed if its extension is to be controlled.

Nævus sometimes attacks the nose or ear, in which case I have seen a peculiar degeneration ensue, involving a destruction not only of the morbid structure, but also of the cartilage of those parts. The upper figure of the accompanying lithograph, shows a nævus

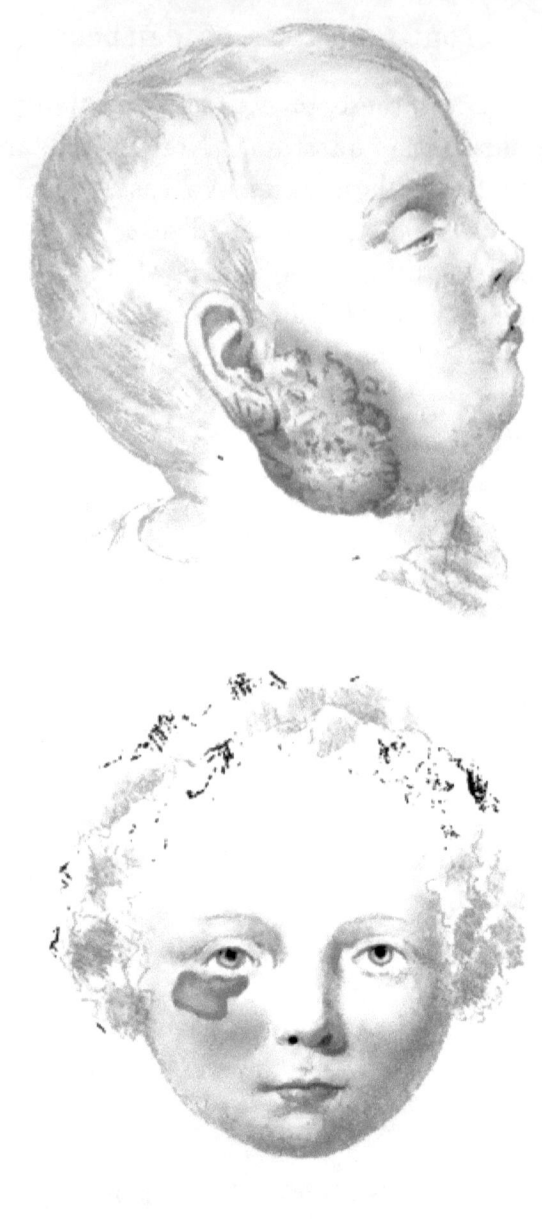

attacking the side of the face, and implicating also the lobe of the ear, destruction of which is commencing. The patient, a little girl aged eight months, was under the care of Mr. Hilton, and the nævus was cured by the use of the electric wire. In one case of this kind, the potassa fusa was applied to the nose, and the destruction of the tissue was attributed to that cause, but, I think, wrongly; for in another case, in which the nævus affected the external ear, though no application whatever was used, the lobe and half the concha disappeared.

Having determined on operating, our aim must be entirely to destroy the affected tissue. So tenacious of vitality are these structures, and so prone are they to extend, that unless their removal be absolutely perfect, they are liable to become almost as bad as ever. The plans that have been adopted for getting rid of them are innumerable; in fact, almost every known irritating or corrosive substance has in turn been called into requisition. To classify the methods of treatment that have been recommended is hardly possible, though attempts have been made by several authors. Perhaps as good a classification as any is that given by Mr. Birkett in the *Guy's Hospital Reports* for 1851, though it includes some measures which I consider wholly inadmissible.

"1. To induce or await atrophy of the new growth by the employment of

    a. Compression.

    b. Astringents or refrigerants.

    c. Ligature of the vessels of supply.

"2. To excite inflammation in the tissue of the nævus, and thus obliterate the cells of the new tissue, by the employment of
    a. Seton.
    b. Acupuncture.
    c. Laceration of the tissue by punctures.
    d. Incision, and the insertion of sponge.
    e. Cauterization with potassa fusa, chloride of zinc, &c.
    f. Injections of stimulating solutions.
    g. Punctures, with a probe coated with nitrate of silver.
    h. Vaccination.
    i. Punctures, with a lancet's point covered with croton oil.

"3. The entire removal of the new growth by
    a. Excision of the disease only.
    b. Amputation of the part affected, as of lips, prepuce, labium pudendi, fingers.
    c. Ligature in various ways.
    d. Complete destruction with caustics."

This classification closely accords with that given by Malgaigne, in his *Operative Surgery.*

The principle by which the *treatment* of nævi is, in all cases, to be guided, is to secure their total annihilation with the least amount of danger, suffering, or subsequent defacement. The means to be employed, therefore, will depend partly upon their seat, whether involving the skin, the areolar tissue, or both; that is, whether cutaneous, subcutaneous, or mixed: partly upon their situation, whether on exposed parts of the body, as the face and neck, or elsewhere; and partly upon the treatment that may have been previously adopted. But although there are various modes of procedure which are specially adapted to nævi of certain forms, or in certain situations, there is one plan which is applicable to all kinds and to almost every place, and that is removal by ligature.

And to this plan, with certain modifications, I give a decided preference. The cases are rare in which other modes of treatment are preferable; and in a majority of instances, no other method affords a good prospect of success.

CHAPTER XIV.

TREATMENT OF NÆVUS.

(Continued.)

I SHALL now describe in detail those plans of treating nævi which seem to be the most valuable, and almost all of which I have myself tried.

1. *Excision.*—This was the plan usually adopted previous to 1825. No wonder, therefore, it has the recommendation of John Bell and of many other men eminent in their day; though they all warn us against cutting *into* the structure under any circumstances. If excision be practised, care must be taken that the knife be applied beyond the circumference of the nævus. In the case of the pendulous variety, this can easily be done. The objection to the operation lies in the fact that the vessels around a growing nævus are the seats of an active circulation, and often are somewhat enlarged; the loss of blood thus becomes unavoidably greater than it is advisable any child should incur. Several ligatures are necessary if the nævus be of any size, and secondary hæmorrhage sometimes follows. Mr. Hilton employs an instrument for compressing the skin in these cases, and prevents hæmorrhage by the actual cautery, leaving the wound to granulate. An objection to this plan is

the unsightly scar that is left. When the actual cautery is not employed, the edges of the wound may be accurately brought together, and thus immediate union be to a considerable degree obtained. This is especially desirable on the exposed parts of the body. The cases to which excision is applicable are—first, pendulous nævi situated in the loose skin about the neck, or in the labia of the female, which latter it is especially necessary should be removed. Secondly, those seated on the lip, where, should there be any hæmorrhage, it is easily controlled; and, in fact, the operation is very similar to that for hare lip, or the removal, by a V-shaped incision, of a malignant growth in an adult—in the accompanying lithograph, nævi of this character are shown. Thirdly, all cases of degenerated nævus, in which extensive hæmorrhage is not likely to take place. And, lastly, any cases of subcutaneous nævus in which the diagnosis may have been so difficult that the surgeon has determined to remove the growth, be it what it may. These are doubtless the cases of "bleeding tumour," in removing which the older surgeons met with the frightful and even fatal hæmorrhage, of which they have left us the accounts. In a case reported by Mr. Wardrop, the patient, an infant, expired during the operation, from the profuse bleeding. I subjoin two or three examples illustrating removal by the knife.

Case 1.—A boy, aged eight months, was admitted into the Royal Infirmary, in March, 1860, with a large nævus upon the neck, occupying about half the space between the chin and the lower jaw on the left side. It was easily lifted up by the fingers from the deeper-seated parts, and the case appeared a very favourable one for excision. Accordingly, I first compressed the skin by the clamp recommended by Mr. Hilton, then cut off the growth, and applied the actual cautery. One vessel, however, seemed to have escaped the iron; for, a few minutes after removing the clamp, a great deal of hæmorrhage ensued. The vessel was tied, and the wound healed perfectly by granulation.

I will next relate a very interesting case of degenerated nævus removed by incision.

Case 2.—E. G., a healthy child, aged four years, was admitted into Guy's Hospital, under my care, in August, 1857. The mother had noticed, soon after birth, a swelling extending around the axillary edge of the left pectoral muscle. This swelling could be emptied of its contents like a sponge, but would refill directly afterwards. It had gradually increased in size, and during the last month had become harder. The mother says she has not been able to squeeze it away. Upon admission, the tumour was the size of a small orange, hard in some parts and soft in others. There was no appearance of enlarged veins in the skin over it; but there was the warty look characteristic of, and, I believe, peculiar to, degenerated nævi. The appearance might be compared to that of a crop of herpes upon the growth. The swelling could be with some difficulty partially detached from the muscle beneath, and appeared mobile. It extended into the axilla. I excised the tumour, in doing which the sheath of the axillary artery was exposed. A considerable portion of skin having necessarily been removed, the edges of the wound were brought as near together as possible, and the rest left to granulate. Great care was needed to prevent undue contraction, and during the process of healing the arm was kept away from the side by an angular splint. On examining the tumour the characteristic cells of nævus tissue were visible; but they were filled with a clear fluid, and thus presented the appearance of small cysts.

The history left no doubt that this was a case of subcutaneous nævus, which from some cause had undergone the cystic degeneration.

Nævi may remain in this condition of cystic degeneration for many years, as is shown in the following case :—

Case 3.—H. B., aged twenty-five, pale and anæmiated, was admitted into Guy's Hospital, under Mr. Birkett's care, in August, 1855, with a large degenerated nævus on the right side of the body. It was of an irregular oval shape, and extended, vertically from the fifth rib to the crest of the ilium, and transversely from within two inches of the umbilicus to about three inches from the spine. It had existed from birth, and had gradually attained its present size. Nine months before, the cutaneous vessels began to exude blood in small quantities. A warty growth then formed upon the surface, and gave rise to a peculiar offensive discharge, which still existed. Bloody serum was constantly oozing from the surface, and there was a good deal of superficial ulceration on the surrounding skin, which was also œdematous. The growth was apparently mobile, and had not lately increased in size. Ice was applied; the ulceration healed, and the œdema subsided, but the oozing of bloody serum still continued. The growth was, therefore, removed by the knife, by means of a large oval incision, and the surface was left to granulate. There was very little hæmorrhage, and the patient regained his health. Upon dissecting off the cellular tissue from the under surface of the tumour, the latter presented the characteristic lobes of nævi, but containing clear fluid instead of blood.

2. *Pressure.*—In one class of cases, that of cutaneous nævi about the scalp, pressure, if assiduously applied, with good counter pressure, is an efficient remedy. And when other plans of treatment cannot be adopted, either on account of the extent of the nævus, or the necessity of avoiding a scar, or from an insuperable dread of the knife or ligature, on the part of the patient's friends, pressure may be tried in other situations; although it cannot be relied on to produce any satisfactory result. It is more powerful when combined with cold, and in this form it was frequently used by Abernethy. He mentions a case of nævus on a child's fore-arm, in which, after six months' pressure by wet bandages, "the skin was pale, and had a slightly shrivelled appearance, and the vessels, which had been like the entrails of a pig, became like solid cords, interposed between it (the skin) and the fascia of the fore-arm."

I believe I have seen instances in which pressure applied unconsciously, by means of the patient's dress, has caused the obliteration of a nævus; when situated over the ribs, for instance, which afford a counter pressure. Perhaps the mysterious disappearance of nævi, which occasionally happens, may be partly accounted for in this manner.

The frontispiece shows a case in which pressure is being applied at the present time, and already with marked effect. In a case of this description, in which the whole limb is involved, it is quite impossible to adopt any other plan.

The modes of applying pressure are various; pads, bandages, strapping, plaster of Paris, have been used, or a kind of internal pressure has been attempted, by passing a cutting needle into the growth, and breaking it up in all directions, and then applying a pad and bandage. This might be a very efficient method if it were not for the difficulty that exists in completely cutting through the structure. The fibrous tissue which bounds the lobes yields before the edge of the instrument. When the nævus is situated about the head, the plan which I prefer, and which I have several times found to be perfectly successful, is the following. A piece of lead a little larger than the nævus is wrapped in a single fold of lint, and to its opposite sides are sewn the ends of an elastic band, not the vulcanized India-rubber, which, owing to the sulphur it contains, ulcerates the skin, but one covered with silk. The plate thus prepared is applied to the part of the head in which the disease is situated, and the elastic band carried round the head so as to keep the plate in position. If on the vertex, the band will be carried beneath the chin; if on the forehead, it will be placed around the back of the head, and so on. Two bands may be necessary to keep the pressure accurately applied. Much depends on the mother's ingenuity and perseverance. I have found six weeks or two months of this treatment suffice to obliterate the vessels; and as the patients have not returned to me, I suppose the effect was permanent. At the same time, I believe the French surgeons are in the habit of keeping up pressure in these cases for one or two years. This method of treatment is more applicable to the true cutaneous nævi. The upper lithograph facing page 210, represents a case in which this plan was successfully applied.

Direct pressure by the mother's finger, applied whenever the opportunity occurred, is said, on the authority of Boyer, to cause obliteration of small nævi on the face. Doubtless, many a mother, anxious to save her child from the interference of the surgeon, has had recourse to this proceeding; and it is pleasant to think that their perseverance may be rewarded. Whether the prescription of *licking* the nævus daily has gained its credit through the same principle, I will not pretend to say.

Covering the nævus with collodion frequently renewed, constitutes a modification of the principle of pressure. In the case of small cutaneous nævi, this plan is often successful.

Pressure has been applied in another way, viz. by passing a suture beneath the nævus, and attaching a piece of bougie to each end, constituting, in fact, a quill suture, which, when drawn tight, serves to cut off the supply of blood. This appears to me a very ingenious plan, but I have not tried it.

3. *Ligature of Vessels.*—With the same view of cutting off the access of blood, the vessels supplying the growth have been tied. This plan has been chiefly adopted in cases of aneurism by anastomosis, where a thrill is communicated to the finger, and the vascular constituents of the morbid structure are altogether on a larger scale than in nævus properly so called. Formerly, it was the practice in the case of large subcutaneous nævi, to tie the main artery of the part; but this idea is now wholly abandoned, and Mr. Prescott Hewett has satisfactorily proved that it is not only unnecessary, but useless, and that the vessels of supply, if any, are the ones which must be tied.

4. *Vaccination.*—In the cases for which it is suitable, this is, perhaps, the most elegant of all the methods which have been devised for the cure of nævus. When the nævus is tolerably small, and is situated in some delicate part—as, for instance, near the inner angle of the eye, on the eyelid, or cartilage of the nose, or on the cheek of a young girl—before any other plan be tried, it may be vaccinated. And sometimes this is remarkably successful. It may be feared lest the Vaccination Act, requiring children to be vaccinated within three months, may tend to do away with this advantage, since nævi often do not present their true characters until long after that age. The difficulties of this practice are, that the child may

have been already vaccinated, in which case the operation may fail to excite a sufficient degree of action to obliterate the tissue; or, in consequence of the tenuity of the skin, which causes the slightest puncture to be attended with a great deal of bleeding, the inoculation may not be effected.

It is necessary that several punctures should be made over the tumour; these should be equidistant from each other, and so near, that the inflammation may extend through all the intervening space, or the cure will not be complete. The pustules should run their ordinary course, and in this case the protection to the system is equal to that afforded by any other mode of vaccination. At the end of about six weeks, the surgeon can judge pretty well whether the cure will be perfect or not. Revaccination may be required, if the pustules do not run their normal course. But if they do, and the cure be only partial, some other plan should be at once adopted.

5. *Irritants and Escharotics.*—These are rightly applicable only to the cutaneous nævi; but they have been employed also in the mixed variety. The chief substances that have been used, and many of which I have myself employed, may be briefly mentioned. The more popular remedies, when the nævi are large, and at the same time superficial, are tincture of iodine and the nitrate of silver in stick. Each of these substances requires to be applied daily, or every other day, by the surgeon, for a period always of many weeks, sometimes of months. They may be considered as the mildest remedies that can be used with any

prospect of success; they demand a great deal of patience and perseverance, and even then are not always successful. Corrosive sublimate and the acid nitrate of mercury have been applied as caustics. Tartar emetic, either in the shape of liniment, ointment, or plaster, has been used to produce pustules, and so to induce obliteration of the nævoid structure; it has been most successful in the form of ointment.

Chloride of zinc, a very powerful escharotic, has been largely used. Formerly it was very unmanageable, but it is now made into sticks, like the potassa fusa, and can be directly applied to any spot. Nitric acid was also largely employed, and bid fair to become an established remedy. It is still somewhat extensively used by those not specially devoted to surgery; but however carefully I have employed it, the result has seldom been satisfactory.

One of the most fashionable remedies a few years ago, for the destruction of nævi, was potassa fusa. Mr. Wardrop, for example, described its results as highly satisfactory, and spoke of it as rendering all other proceedings unnecessary. It has also been much used abroad. Chelius says (South's translation), "It is, in all cases, to be considered as the most proper, when the swelling is broad and superficial, especially in children." He also observes: "If on the first cauterization the disease be not entirely destroyed, and show itself afresh, I have never noticed its quicker spreading, and it is always cured by repeated cauterization." One very simple way of applying it is, by frequently touching the nævus, just sufficiently to produce a superficial slough.

Mr. Lawrence considers that the use of potassa fusa in large nævi would be as dangerous a practice as excision; and Boyer states, that he has seen death result from the treatment. My own experience has led me to the conclusion, that the caustic potash may be safely and advantageously employed in superficial nævi, which involve a great breadth of surface, and which cannot be treated by ligature or excision. For instance, if a nævus be situated at the lower angle of the nose, involving portions both of the nose and cheek, so that neither the knife nor the ligature can be applied, the caustic potash is the best remedy. Generally, more than one application is required. The mode of proceeding is, first to cover the surrounding skin with plaster, to guard it thoroughly from the effect of the deliquescent potash. The stick should then be well rubbed over the nævus, being carefully applied to every part of the surface. A dry scab forms, which at the end of ten days or a fortnight, will drop off, leaving a healthy granulating surface beneath. If any part of the nævus structure remains, the caustic must be again applied. It is necessary to destroy a narrow margin of skin beyond the apparent boundary of the nævus. On the Continent, potash and lime, in the form of the Vienna paste, are largely employed.

In addition to the escharotics that I have mentioned, arsenic, hydrochloric acid, sulphuric acid, &c., have been used. Of these, arsenic, of course, is inadmissible, on account of the danger of absorption; hydrochloric acid is useless, and sulphuric acid painful and unmanageable.

6. *Introduction of Caustic into the Substance of the Nævus.*—This has been recommended by Sir Benjamin Brodie, who calls it subcutaneous cauterization. He first divides the fibrous tissues of the nævus, by means of a tenotomy knife passed beneath the skin, and then introduces a probe, armed with the nitrate of silver, which he moves in every direction through the growth, so as to induce inflammation and obliteration.

Sir Benjamin remarks, " I have used this on several occasions with great advantage, especially when the tumour has been on the face, where it was a great object not to destroy the skin. I was requested to see a little child that had one of these subcutaneous nævi at the tip of the nose, giving it a very ugly appearance. By far the greater part of the alæ of the nose was involved in the tumour, and to have cut it out would have disfigured the child for life. I treated it according to the method which I have just explained; several operations were required, but they succeeded perfectly. The child is quite cured of the nævus, and I will not say that you see no mark at the end of the nose, but there is so little, that unless your attention were called to it, you would not know that anything had happened." Sometimes, however, ulceration of the skin follows this treatment, and other plans are more adopted at the present day.

7. *The Seton.*—The use of a seton is a very old plan of treating nævus. Mr. Fawdrington, of Manchester,

seems to have adopted it very extensively. And it still has advocates. One of its asserted advantages is, that it is followed by very little disfigurement, there being no other scars than those produced by the passage of the needle. The seton is applied in various ways; sometimes a single thread is passed through the growth; sometimes several. Mr. Erichsen recommends that a number of fine silk threads should be introduced in different directions, and left for a week or two until they have set up a sufficient amount of adhesive inflammation; the remaining parts of the tumour are then to be treated in the same manner. In this way its consolidation may be gradually effected. Mr. Curling also speaks highly of setons; he passes three or four, containing two threads each. The needle should be introduced through the sound skin around the nævus.

When a seton is used, care should be taken that the thread employed is larger than the needle, so that it may plug the openings in the skin, and prevent hæmorrhage. Should it not produce sufficient inflammation, irritants, such as tartar emetic ointment, or castor oil, may be smeared on the threads, and they should be kept in long enough to induce suppuration. Setons, of course, are not applicable to the cutaneous form of nævi.

8. *Heated Wires and Actual Cautery.*—I have had but little experience of either of these plans. I tried the former in one case of nævus on the cheek, but found great difficulty in keeping the needles hot. The consequence was, that while they made a larger

aperture in the skin than was desirable, they had little effect on the structure of the nævus. The needles were heated by a spirit-lamp. The electric cautery is more convenient and successful. As regards the actual cautery, Dupuytren is the only surgeon that I am aware of who has used it much. It is a very effectual method, but extremely difficult to apply to exactly the right extent.

9. *Needles and Twisted Sutures.*—Dr. Marshall Hall recommended that a " needle, appropriate in size to the magnitude of the nævus, should be taken and inserted at one point of the circumference of the nævus, and be carried through the substance of the disease from that point to various other points." He says, " Each of these movements of the needle induces an incision, which being treated as adhesive inflammation, substitutes cicatrix for the morbid vascular tissue, and so cures the disease in a certain degree." This operation, which he termed acupuncture, would have to be repeated at regular intervals; it may be several months before the cicatrization is complete; but Dr. Marshall Hall considers time nothing, compared to the horrid plans of treatment usually practised—the scalpel, ligature, or caustic.

Lallemand's practice was to pass needles into these growths, and in one case he used as many as 120. He also used to pass long needles through the swelling, bend the points up, and let them remain until they suppurated out. The nævi turned black from effused blood, and became absorbed. If any swelling still continued, he renewed the application.

Twisted sutures have had their advocates: Keate, Brodie, White, and others. The tumour is obliterated by strangulation. Needles are passed beneath it, and a twisted suture applied over the whole. A partial sloughing occurs, which, however, may or may not be attended with the cure of the nævus. This plan is chiefly adapted for the large pulsating tumours, which are sometimes, though rarely, seen over the cranial bones of children.

In the *Lancet* for October, 1850, Mr. Ferguson recommends a combination of the needle and twisted suture with pressure. He advises two strong pins to be passed through the base of the swelling in one direction, and two others transversely to them. A piece of lint, rolled up, is to be applied under the needles around the tumour, a pledget of lint laid on the nævus, and threads twisted around the pins, so as to apply pressure over the lint on the diseased structure. The plan of treatment is applicable to all forms of nævi, but is hardly to be recommended for any.

10. *Injections.*— In the *Medical Gazette* for October 1st, 1836, is a very admirable paper by Mr. Lloyd, entitled " Observations on the Treatment of Vascular Nævi Materni," in which the use of injections is for the first time advocated. Mr. Lloyd used a small syringe fitted with tubes of different sizes. The point of the tube was introduced through the skin at a little distance from the growth, and various fluids injected; nitric ether; nitric acid (6 drops to ʒj of water); solutions of chloride of lime, sulphate or acetate of

zinc, hydrochlorate of ammonia, aromatic spirit of ammonia, or iodide of potassium. Other substances were soon added to the list. Mr. Lloyd considered injection to be "applicable to nævi so large or so situated as to be wholly irremediable by any other means." This opinion has not been borne out by facts; but nevertheless the practice thus introduced has proved to be of great and permanent value. To a modification of it, I give the preference in the true subcutaneous nævi; and these are the only cases in which I think it should be adopted. I believe that it has sometimes fallen into undeserved ill-repute through want of a discrimination of the cases to which it is applicable.

Injection of nævi has been attended with fatal results. In 1837, the year after the first introduction of the practice, a child was treated in the Leicester General Infirmary for a nævus over the maxilla; two injections had previously been employed, the first of nitric acid, the second of nitric ether. The third time ammonia was used. Whilst the injection was going on the child died. This case is reported in the *Medical Gazette* for Dec., 1837.

Until the recent introduction of the perchloride of iron into surgical practice by Malgaigne, the operation of injection for the cure of nævi was comparatively unsatisfactory. This substance, however, if used in certain picked cases, is so efficient, and with proper precautions is so safe, that it has, in my opinion, given the procedure by injection a place among the best remedies for this affection. The cases, however, must be carefully selected, or disappointment will ensue. When the nævus is of the true subcutaneous type, the skin not being involved, and when it is situated in a conspicuous part of the face, neck, or arms, where a scar is to be avoided, injection of the perchloride of iron is the plan I always adopt. But these cases, it must be remembered, are not frequent.

The mode in which I perform this operation is as follows:—First introducing through the skin a narrow, double-edged knife (represented in the accompanying woodcut) about half an inch from the apparent margin of the nævus, I pass it into the substance of the growth, and move it in all directions, to divide the tissue. If the nævus be large, I do this from several points. As the knife is withdrawn there is a flow of blood, which the pressure of the finger will control. Then, the syringe being previously charged with the solution of perchloride of iron, I introduce the point of it into the aperture, or into each aperture, in turn, if there are more than one, and throw into each opening about four or five minims. Coagulation of the blood instantly takes place; and the tumour becomes hard and prominent. Absorption gradually ensues, and the structure is destroyed. If the injection penetrate the whole nævus, as evidenced by the hardening of the whole, no further operative procedure will be necessary; if, on the contrary, any portion retain its softness, that part should be injected through a fresh opening, as soon as convenient. When the absorption is complete, no trace is left to show where the disease has been; but this process may occupy several months, or even a year.

The instruments I use are figured in the accompanying woodcuts. The syringe is of glass, with an accurately adapted piston. It terminates in a narrow platinum tube, in one side of which, near its extremity, is an oval aperture; over this tube, a second tube is placed, terminating in a trocar point, and having a similar aperture. By rotating this outer tube, the openings may be made to correspond, or not, at the will of the operator; the open or shut condition being marked by the indicator pointing to the letter O or S. The tube, of course, is introduced closed, and opened when within the growth. There is a modification of this instrument in which, by turning a screw, a single minim of the perchloride can be injected at a time; but I do not use it.

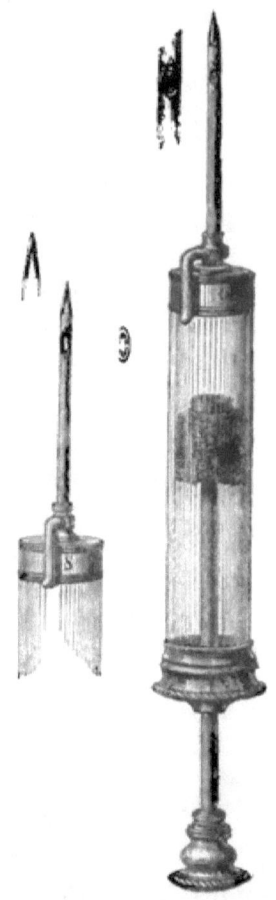

The preparation of the perchloride that I use is made for me by Mr. Hodge, of the Borough. If too much be thrown in, it appears to act as an irritant, and suppuration or sloughing may occur. I have seen in one case severe convulsions immediately follow the injection, but the child soon recovered. I am by no means sure, however, that any other mode of interference with the nævus in this case would not have had the same effect.

Case 1.—A. N., aged five months, came under my care in February, 1857, on account of a subcutaneous nævus behind the angle of the jaw, on the right side. It was the size of a large chestnut, and was increasing rapidly. It could be entirely emptied by pressure. I injected it in three places with the perchloride of iron; the whole tumour became hard, and required no further interference. Slight suppuration from one spot occurred a fortnight afterwards. The swelling was eight months before it entirely disappeared. I saw this child a month ago, and there was then scarcely the slightest indication that the nævus ever existed.

Case 2.—L. C., aged eleven years, was admitted into the Infirmary for Children with a subcutaneous nævus, the size of a small egg, on the right side of the neck, behind the border of the sterno-mastoid muscle. Six years previously it had been treated by ligature, which had evidently failed. Free division of the structure was made, and the perchloride was injected in two places, ten minims in each. For three or four days she complained of some pain; but there was neither constitutional disturbance nor inflammation. The induration gradually subsided, in the course of about six weeks. She is now quite well, and with scarcely a scar.

Case 3.—E. C., a child aged twelve months, was brought to the Infirmary for Children with a subcutaneous nævus on the vertex of the head. The skin, in this case, was partly implicated. I injected the tumour with the perchloride of iron. In the course of a week the unsound portion of skin sloughed. The cure was perfect, but there remained a small scar.

It was this case which first led me to a clear recognition of the form of nævus for which injection is the appropriate treatment. When the skin is involved in the disease, we can expect only one of two results; either that it will slough, or that the cure will be imperfect.

While on this subject I may mention, for those who attach importance to such ideas, that Dr. Brainard, of America, recommends the lactate of iron as a substitute for the perchloride, on the ground that the elements of the former are normal constituents of the blood.

11. *Ligature.*—The last mode of treatment that I have to describe may be considered as, upon the whole, the most satisfactory. It is more easily applicable than any other to all forms of the disease, and a needle, pins, and some strong four-twist thread, are all the apparatus that is required. There are, however, a great variety of modes of ligature, which are appropriate respectively to various cases. The chief of these I will describe and illustrate by examples.

a. The simplest way of applying the ligature is to

pass two pins, at right angles to each other, under the nævus, and encircle the whole with a ligature of four-twist thread. This should be tied in a bow knot so tightly as entirely to strangulate the part; the

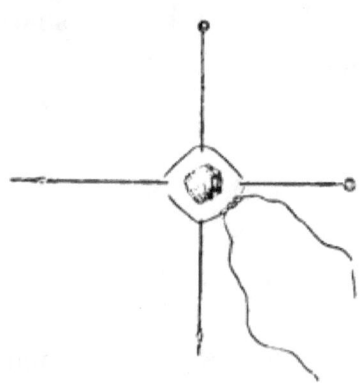

pins should then be removed. At the end of four hours I untie the knot and remove the ligature. This plan is suitable for very small nævi, of the mixed or cutaneous kind. The ligature is seen in the accompanying woodcut.

Case.—The way in which I first lighted on this plan was by an accident. A child was sent to me by my friend Mr. Roper, in June, 1857, with a nævus of the mixed variety, about the size of a fourpenny piece. I adopted the treatment which I was in the habit of using for small nævi; viz., passing two pins at right angles to each other under the mass, placing one ligature round the whole, tying it tightly, and withdrawing the pins. My design was, that the ligature should continue on the nævus, the effect of which I knew to be that, in the course of four or five days, the latter would shrivel up and drop off with the thread. To my annoyance, however, the ligature came off accidentally four hours after it was applied. I feared that probably the usual result would not be obtained; but was agreeably surprised to find that the nævus, which had been tied, dried up and formed a shrivelled mass, under which the curative process went on without any suppuration; when it dropped off (which it did in twelve days), there was scarcely any scar to be seen.

Guided by this lucky misadventure I have since, in small nævi, adopted this plan of treatment. I find that the nævus dries up and forms a kind of scab, which separates at the end of fourteen days or so, leaving a healed surface underneath. The rationale of the treatment appears to me to be, that the vessels compressed by the ligature have the circulation in them arrested for a sufficient time to allow the blood to become consolidated; but are not entirely obliterated so as to cause the parts to slough; consequently the strangulated mass becomes atrophied, and is thrown off without inflammation by the healthy tissue beneath. The scar on this plan being less than when other forms of ligature are employed, it is of course especially adapted for nævus on the face.

b. Another simple mode of ligature is, to pass a needle, armed with a double thread, under the nævus, the points of entrance and exit being at least a quarter of an inch beyond the boundary of the altered texture. 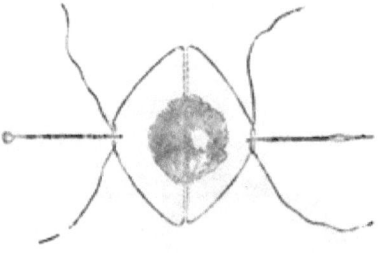 A pin should then be also passed underneath the nævus transversely to the threads, and each half of the nævus should be separately tied by one of the threads beneath the pin. The pin may then be withdrawn, or its ends cut off short, at the operator's pleasure. Let me observe here, that in all cases, except those in which the ligature is to be shortly removed as above described, nævi cannot be tied too tightly. Not only is this desirable in order to secure the destruction of the part, but also on account of the less amount of suffering which a very tight ligature inflicts. When the utmost tension is employed, the little patients, on awaking from the chloroform, often seem to experience scarcely more pain than they might from an incised wound.

When the nævus is thus encircled by the two threads, it becomes tense and swollen. Its surface grows livid, and exudes a straw-coloured fluid, apparently the serum of the blood. The presence of this exudation proves that the ligature is perfect. Some think it desirable to puncture the nævus at this stage; but I do not myself adopt this plan, nor do I think any benefit results. The mass begins to shrivel up at the end of twenty-four hours, and after five or six days, the threads having ulcerated their way through, it can easily be detached, leaving a healthy granulating surface. In the course of a few days these granulations are apt to become particularly exuberant and flabby, and will sometimes give a great deal of trouble. The usual treatment, dry lint, nitrate of silver, &c., may be resorted to. Of course a large white cicatrix remains; and if the ligature have not completely encircled the nævus, a small red spot may be observed at one angle of it, suggesting to the surgeon the unpleasant probability of a return of the affection.

This method of ligature is suitable for mixed and cutaneous nævi, of a somewhat larger size than those before referred to.

Case.—L. W., aged three months, was brought to me in July, 1857, with a "mixed" nævus on the forehead, of elongated form, and situated rather on the left side of the mesian line. It was three-quarters of an inch in length, and was rapidly increasing. As it was elongated and narrow in shape, I thought it a fit case for this form of ligature. The threads were passed beneath the growth, in the direction of its long axis, so as to include by each a narrow strip, and tied around the pin. In about a week, the dried mass was removed, and the surface granulated healthily.

c. Two needles, each armed with double thread, may be passed, at right angles to each other, beneath the nævus, and then the needles being cut off, the four threads should be tied to one another, so as to enclose the structure. This, of course, is effected by tying one thread of each ligature to one of the other. 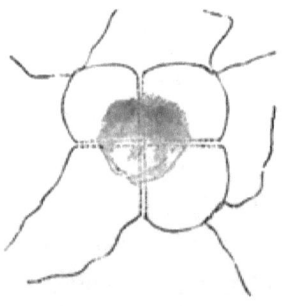 The first three knots must be tied somewhat loosely, and the whole is to be tightened by drawing firmly upon the fourth. By some surgeons, the skin is divided in the course of each of the ligatures, with the view of diminishing the pain. I am not in the habit of doing this; there is

risk of cutting the thread, and the hæmorrhage, though not great, may be inconvenient. When the nævus is wholly subcutaneous, and not too large, a plan has been adopted of reflecting the skin from it by means of a crucial incision, and placing the ligature beneath. I have seen this practice successfully carried out, but the hæmorrhage is very considerable.

The advantage is, of course, that the skin is saved, and the ligature is applied directly to the disease.

There is a simplification of this form of ligature which I have used for many years. One needle only is employed, and it is threaded a second time with the ends of the threads which have already been carried beneath the nævus, and then is passed at right angles to its former course. By this means, the thread is carried half round the tumour, and two knots are sufficient instead of four. My present method of procedure, is this: Taking the needle depicted in the accompanying woodcut, I pass it, with the slit closed, under the nævus, entering nearly half an inch from its margin, and emerging at the same distance on the opposite side. I then open the slit, introduce into it a loop of thread, and fixing it, by closing the slit, draw it beneath the nævus. Having done this, I cut one thread of the loop, and carry the other, which remains in the needle, a quarter round the nævus. The needle is then passed at right angles to its former course, and the thread liberated. The end of the other thread is then to be taken up by the needle, in the act of withdrawing which it will be pulled through to the other side. By this means, there remain only four ends, which are to be tied two and two, so as to describe a figure of eight; and the whole mass is strangulated, as shown in the accompanying woodcut.

Mr. Ferguson has described, in the *Edinburgh Medical Journal* for 1847, a ligature somewhat on this principle. He, however, puts two different threads through the eye of the needle at the same time. The difficulty of passing the second thread through the same eye is avoided by the plan described above.

If tied sufficiently tight, this ligature separates the nævus in about the same time that the others occupy, viz., six or eight days. It leaves a granulating sore, which contracts as it heals, forming an irregular cicatrix. The best application is cold-water dressing. The cases suitable for this treatment are extremely frequent. An "out-patient day" at any large hospital seldom passes without presenting one or more.

d. If the nævus be very large, it is quite impossible, of course, to include it all in the last-described form of ligature. Some other plan, therefore, must be adopted. One suitable for many cases, is to take a needle, armed with a double thread, and pass it repeatedly across beneath the nævus, beginning about a quarter of an inch from one extremity; in fact, to underlay it, as it were, with a double uninterrupted suture. A loop of four or five inches of the thread must be left (for tying) beyond the border of the nævus, at each turn. And then one thread of

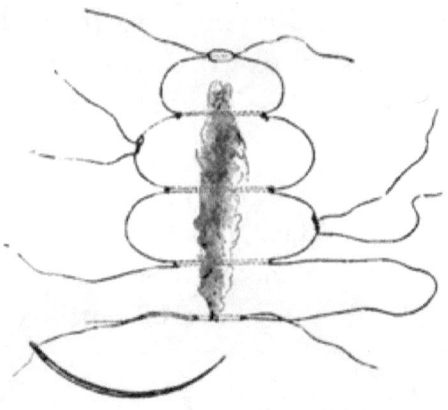

each of these loops must be cut, and the threads from the adjacent loops tightly tied, so as to strangulate the included mass. The accompanying woodcut will render clear the method of proceeding. Considerable care is needed to avoid cutting the wrong thread.

The rule is to cut, on each side, always the same thread; and it has been suggested to colour one thread black, and to cut the black on one side and the white on the other. This is, doubtless, a safe guide; but pulling on the threads at the time of division is the plan to which I have recourse. The subsequent progress of the case is the same as in other forms of ligature.

e. Another ligature, similar to the last mentioned, but having some advantages, has been introduced by Mr. Curling. He uses a needle fixed upon a handle, with which he passes a loop of thread beneath the nævus, near one extremity. The needle is then withdrawn, sliding on the thread; and, at a short distance, is introduced again, carrying with it another loop of the thread. This is repeated as often as necessary; one thread sufficing for the whole. The loops left by withdrawing the needle are then divided, and the threads of the adjacent loops tied; the terminal portions of the thread being tied around the end of the nævus. By this plan, all the knots are on one side; and there is no risk of cutting the wrong thread.

In using this ligature, the slit-eyed needle, depicted on p. 242, would be found convenient.

In addition to the modes already described, numerous other plans of ligaturing nævi have been suggested by various surgeons, and it is probable that many more will be suggested hereafter, since the operation affords a boundless scope for ingenuity. Those that have been mentioned, however, are the simplest and most efficient, and embody the principles which must be applied in every case. I shall proceed to describe another mode of treatment, also by ligature, which differs from the foregoing in an essential point.

12. *Subcutaneous Ligature.*—This consists in encircling the nævus with a thread passed beneath the skin. There are two plans by which this may be accomplished, according to the size of the nævus. The first mode is to use a needle describing a large curve, and with it to carry the ligature subcutaneously half round the nævus. The needle is then to pierce the skin, to be re-introduced at the point of exit, and to be carried round the other half of the growth to the original point of entrance. By tying the two ends of the thread at this point, strangulation will be effected. Or the needle may pierce the skin at three or more points in the circumference of the nævus, according to the circumstances. The dotted lines in the upper of the accompanying woodcuts indicate the course of the ligature. It is necessary, of course, to carry the needle to a sufficient depth to include the whole diameter of the growth.

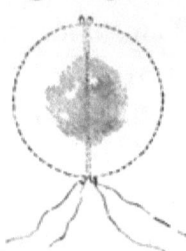

Another form of the subcutaneous ligature is shown in the lower of the woodcuts. The curved needle, armed with a single thread, is passed subcutaneously half round the nævus, and then brought out through the skin as in the former case. At this point the operations differ; the needle, on being re-introduced at the point of exit, instead of being carried round the other half of the nævus, is passed transversely across beneath it, and the thread brought out at the original point of entrance. By tying the two ends of this thread, one half of the nævus is strangulated. The other half of the nævus is to be treated in the same way, the thread being carried across it in the same track. Care is necessary that the whole mass be included, and that a central portion do not escape the pressure. These very ingenious methods of treating subcutaneous nævi are described at length, and illustrated with cases, in Mr. Curling's paper in the *Medical Gazette* for January, 1850. But they are both of them difficult and uncertain, and they are available only for the strictly subcutaneous nævi, for which injection with perchloride of iron is so admirably adapted. I therefore never have recourse to them.

Now and then it happens that large nævi, involving a great extent of surface, become atrophied, and in process of time almost disappear. Curiously, the more severe the case, the more likely is this to occur; nævi, extending over a whole limb, will very frequently die almost entirely away. In this process they either may or may not pass through the state of cystic degeneration to which reference has previously been made. When this spontaneous cure takes place, it generally leaves a whitish opaque appearance of the skin, resembling the cicatrix occasioned by ligature. In fact, in whatever way nævi are got rid of, there invariably remains some indication of their previous existence. The question arises whether it is even worth while to wait for the chance of spontaneous cure. If the nævus do not increase in size, there is no reason, as I have previously observed, why it should be interfered with. And should the child suffer from any depressing disease, or be subjected to influences which lower its vital powers, the growth may, as all surgeons must have seen, disappear. But if there be the slightest tendency to increase, it is never advisable to delay an operation.

Large nævi also sometimes spontaneously slough, in which case very serious hæmorrhage may occur. This may be restrained by lint dipped in the perchloride of iron. Sloughing commences at the centre, and will frequently extend over a large part of the nævus, and then cease, leaving the marginal portions undestroyed, or even spreading, and requiring operative treatment. This tendency to slough seems, at first, curiously at variance with the fact before mentioned, that often it is very difficult to destroy these growths, and that almost anything may be done to them without impairing their vitality; but in this respect they resemble many morbid formations.

Moles.—One word may be desirable with reference to these deformities, since in some cases, when they are in a conspicuous part of the body, it is desirable to get rid of them in early life. The best method is the

Day & Son Lith^{rs} to The

application of nitric acid. The lower figure in the accompanying lithograph shows a large hairy nævus, or mole, which occurred in a child eight weeks old, who was admitted into Guy's Hospital under Mr. Birkett's care.

CHAPTER XV.

INJURIES AND DISEASES OF BONES.

I. FRACTURES.

THE symptoms and the treatment of fracture differ but little in the child and the adult. They will not, therefore, be spoken of in detail, but a few general remarks are called for.

There is a peculiar kind of fracture in children, known as the "*green-stick fracture.*" This name well characterizes the affection, which is partly a breaking and partly a bending of the bone, like that which occurs in a piece of fresh wood. It is attended with great deformity, presenting an appearance like that of rickets; but from this affection it is easily distinguished, not only by the history, which is always that of fracture, but also by the pain which is present. Unlike rickets, it occurs most frequently in the bones of the upper extremity, and is confined to one limb. The treatment is similar to that of ordinary fracture. The splints must be kept on for a fortnight or three weeks, and special care should be taken that the joints above and below the fracture be kept in absolute rest.

Another form of fracture peculiar to early life is the *separation of the epiphyses.* In appearance this greatly resembles dislocation, and requires careful discrimination. The diagnosis and treatment are the same as in the case of fracture near a joint in the adult.

In the treatment of fracture in children, leather or millboard, well padded with cotton wool, is, in almost all cases, much preferable to hard wooden splints. We cannot too often call to mind the delicate structures we have to do with, and how soon any undue pressure by tight bandaging, &c., may give rise to sloughing. I have seen incipient gangrene of the foot of a child occur within forty-eight hours after bandaging for fracture of the thigh. It is not enough to place a bandage loosely on the limb at the time, allowance must be made for the swelling that arises, and the condition of the limb must be carefully watched. In the first stage of the treatment, sand-bags instead of splints are for this reason very advisable; the application of the splints being deferred until the swelling has subsided.

Respecting *Fracture of the Clavicle*, on account of its great frequency in children, one word may be said. Many parents are very much alarmed at seeing the swelling which this accident produces; but it always does well, and frequently, as I have seen, when no treatment whatever has been adopted. At the same time, it is better to place the bone in the position recommended for fractured clavicle in the adult. I believe, in these cases, the fracture is often of the "green stick" kind.

When *Compound Fracture*, either in the centre of a

long bone, or extending into a joint, occurs in children, the attempt should invariably be made to save the member. Secondary amputations are by no means so fatal as was formerly believed; and it is impossible to say when the injury to the vessels is so great as to render sloughing of the limb inevitable.

II. DISLOCATIONS.

Dislocations are comparatively rare in children, separation of the epiphysis taking place instead. The most frequent dislocation is that of the elbow joint; and owing to the deformity it produces, the diagnosis is easy. The projection backwards of the prominent olecranon, which is the most usual form of displacement, characterizes it at once. Here chloroform comes to our aid; under its influence the bones are returned with the greatest ease.

Compound Dislocations sometimes occur in children, and the question of amputation arises. My own experience is that, even in these severe accidents, so great are the reparative powers in children, the limb may almost always be saved. The attempt should certainly be made. One of the most severe cases that I ever saw was admitted into Guy's Hospital, under Mr. Morgan's care, in 1842. A child, five years old, had met with compound dislocation of the knee-joint. So serious was the injury that amputation was decided on, but the boy was too ill to bear the operation. The dislocated bone, therefore, was reduced; in time, perfect anchylosis occurred, and the child recovered, with a stiff joint.

Compound dislocation of the fingers, for the most part attended with crushing of the soft parts, may occur in children. Under these circumstances attempts should be made, by careful adaptation, and appropriate treatment by splints, &c., to preserve the whole member, or at least to allow nature to perform the necessary amputation.

It does not come within my design to describe this class of accidents in detail. The foregoing remarks comprise that which I have observed as specially applicable to children. Of course, in boys approaching the age of puberty, when they begin to be employed in various occupations, injuries of this kind become more frequent; but they present few peculiar features, and the chief reference of my observations is to children of a younger age. The principles of conservative surgery should, in all cases, be carried to their utmost extent.

III. DISEASES OF BONES.

In treating this subject, my remarks will be limited to the two chief affections of the osseous system met with in children—caries and rickets.

Caries.—A child is brought to us with a swelling upon the instep, or over one of the metacarpal bones, or phalanges. In one case there may merely be a swelling, with an obscure sense of fluctuation, and the child may be said to have been in this state for some weeks without any change occurring; in another case the surface is reddened, but without much tenderness, and the mother says the child has rheumatism. This condition exists for weeks and months. Poultices are applied with no effect. At length the surgeon, feeling satisfied that matter has formed, plunges a lancet into the swelling, and, in some cases, there appears a thin ichorous discharge; in others, a more healthy pus. If, however, the case be left to itself and the abscess burst, an open sore is the result, in consequence of a considerable extent of skin being destroyed. This sore is some time before it contracts, and then leaves a fistulous opening. On passing a

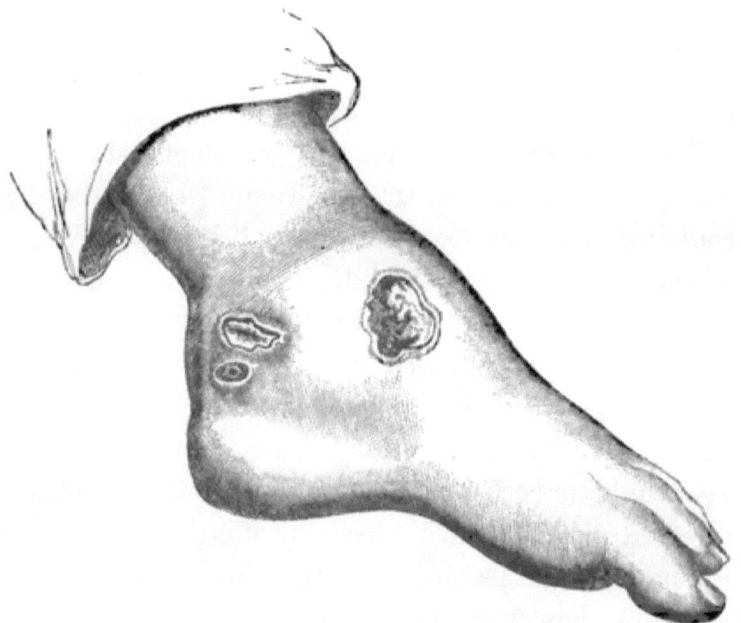

probe along the sinus, the true state of the case is at once revealed, as the probe passes into soft disintegrated bone. The disease, in fact, is ulceration of bone, or caries.

This class of cases occurs only in delicate strumous

children; and the disease may exist for years, affecting various parts, but generally the bones of the hand or foot. If it attack the long bones, it is confined to their extremities; sometimes it seems to spread into the adjacent joints, giving rise to secondary disease of those organs. As the disease progresses, small pieces of bone are thrown off, or the greater part of an entire phalanx may come away. Should the patient be placed in very favourable circumstances, it may recover with a shortened finger or toe, or adhesion of the skin to the edges of the bone, giving a peculiar puckered appearance, indicative of the disease that has occurred.

The best *Treatment* that can be adopted in these cases, is to devote great attention to maintain the constitutional vigour of the child. I do not know any disease in which surgical interference is less required from beginning to end, and yet in which the temptation to interfere is greater. Sadly disappointed would any surgeon be who should lay open the sinuses in the expectation of removing a mass of dead bone. He would only break up a quantity of soft, disintegrated tissue, which bleeds profusely, weakening the patient, and leaving the sore no better than before. There is, however, one exception to this rule; and that is, when the caries attacks the posterior portion of the os calcis. In a case of this kind, I have removed a large mass of carious bone, probably the whole of one centre of ossification, with the best result.

By means of sea-air and generous diet, many a case

that might have appeared hopeless, has got perfectly well, and many a member that has been condemned, been saved.

Rachitis.—This disease, being essentially a constitutional affection, comes as frequently under the care of the physician as of the surgeon. It never occurs after the age of puberty, and appears to be a disease of the nutritive process, the bones being deficient in earthy matter. Rickets chiefly attacks the lower extremities, but the children are usually large headed, as if the contents of the cranial cavity expanded the unresisting bones. The tibia and femur are bent forwards and outwards, and the little creature waddles about in a way distressing to see. The anxiety of friends is to know whether the bent bone can be restored to its normal shape. Of one thing I am quite sure; that no mechanical contrivance yet invented is of any service. I have put splints on the outside and the inside of the leg, and know them to be useless. Boots, irons, &c., do nothing but add to the weight which already overburdens the feeble limb. The only rational mode of seeking to benefit the case is by good hygiene, especially milk and country air. It seems reasonable to add phosphoric acid or phosphate of lime to the food, or in the case of infants at the breast, to administer it to the mother; but in my own experience, I have not been able to decide whether these or the hygienic measures have conferred the benefit. Cod-liver oil and syrup of the iodide of iron also are remedies which I always employ.

In respect to the recovery of the symmetry of the

limb, the bent bone certainly never becomes straight; but as the child grows, and the bones enlarge and lengthen, their tendency is to assume the natural position, and the deformity diminishes.

As a kind of doubtful ground between disease of bone and disease of a joint, we have the condition called *Genu valgum*, or knock-knee. It appears to be connected with relaxation of the internal lateral ligament of the joint; but it seems also, in many cases, to be accompanied with more or less deformity of the upper part of the tibia. These children are generally found to have a large tumid abdomen, which was formerly ascribed to mesenteric disease. The truth seems to be, that they are not sufficiently nourished. The same condition of relaxation very frequently occurs in the ankle joint, with a similar constitutional debility. The children, in both cases, are unable to walk with comfort, or soon become tired.

Treatment.—Under no circumstances should splints or irons be applied, for the reasons before stated. Strapping, however, gives great relief, and frequently enables the patients to take exercise. It should be applied once or twice a week. No other local measures are of importance, though bathing, friction, &c., may be enforced with probable advantage. Constitutional treatment, cod-liver oil and iron, are the main remedies.

CHAPTER XVI.

INJURIES AND DISEASES OF THE JOINTS.

I. INJURIES TO THE JOINTS.

IN addition to the remarks I have made on this subject, under the head of Dislocation and Fracture, I need refer only to *Wounds of the Joints*. These arise from various causes: falling on a spike, kneeling on a nail or piece of glass, &c.; but in whatever way they are produced, the prognosis is much more favourable than in the case of adults. Keeping the parts absolutely at rest is the great requisite. A splint is essential; and the limb should be placed in the position most convenient for after use, in case of anchylosis. The elbow, therefore, should be put up bent at a right angle, the hip or knee, straight. Besides this, cold water or ice may be applied. These applications are very soothing; and may be continued until the inflammatory symptoms are subdued. Should there be reason to believe that suppuration has taken place, the plan, first brought to my notice by Mr. Key, of making a free incision into the joint, should be adopted. By this means, a good chance of anchylosis is attained. Should the disease prove ultimately intractable, resection in the case of the elbow or hip, or amputation in the case of the knee, can but be resorted to at last.

II. DISEASES OF THE JOINTS.

Diseases of the joints in children correspond in great measure to those which are met with in adults. But there are certain forms to which there is a special liability in early life, such are the various affections known under the name of strumous disease, whether affecting the synovial membrane, the cartilages, or the articular portions of the bones. Systematic treatises on the joints describe in detail the symptoms, pathology, and classification of the diseases pertaining to these structures; and into these subjects, therefore, I shall not enter. No peculiarities in these respects have yet been observed in children; but the course of the diseases is different. The frequency of an unfortunate issue of these affections in adults is well known; in children, if the circumstances be at all favourable, and the treatment can be efficiently carried out, the prognosis is, on the other hand, very favourable. The symptoms may be extremely severe, but a termination in recovery more or less complete may be hopefully anticipated.

I shall accordingly describe the diseases most frequently met with in the knee, hip, and elbow joints; in doing which the principles which should guide us in every case will be sufficiently indicated.

1. *Diseases of the Knee.*—A little child is brought to us with the knee-joint swollen, semi-flexed; not painful,

but yet not able to be perfectly flexed or extended without pain. Fluid is evidently contained within the capsular ligament, and extends in front of the lower portion of the femur above, and on either side of the ligamentum patellæ below. The fluid can by pressure be made to pass from the lower to the upper part of the joint, or *vice versâ*, and the swelling over the femur extends more upon the inner than the outer side, owing to the density of the fascia lata. There is not any redness or increased heat, and we evidently have a case of chronic synovitis. The child appears to suffer but little constitutionally, and may be strong and hearty.

Or again, a child may be brought to us with a similar enlargement of the joint, though rather more irregular in outline; there is greater impairment of motion, and instead of distinct fluctuation, a solid tumefaction is felt, resembling enlargement of the bones. There is more or less pain, which can be localized in particular spots, and is aggravated at night; the child seems to suffer a little in health. The case is one of disease of the synovial membrane, extending to the cartilage, and the swelling arises not from enlargement of the bone (a condition which is very rarely met with), but from effusion of fibrine. This effusion serves to keep the joint motionless, being thus of great service, and considered as a remedial process might pass for one of nature's happiest efforts.

These are the most common forms in which disease of the knee is presented to us in the child. Severer forms may occur, but they arise either from the progress of these, or from accidents, or as secondary results from other diseases, such as the opening of an abscess into the joint, &c.

In the *Treatment* of these diseases, the first and most important condition, without which any good result is hopeless, is to keep the limb perfectly at rest. This is to be accomplished by means of a splint, without which it is useless to attempt to confine the child to bed. The best splint is a piece of leather, about three-eighths of an inch thick, well soaked in boiling water. This should be accurately moulded to the limb, padded with cotton wool, and retained *in situ* by a bandage. In the wealthier class of patients some of the beautiful appliances for keeping joints at rest, contrived by Mr. Heather Bigg, may be used.

The splint is an essential condition of treatment in all cases; but in other respects the two forms of disease above described are amenable to different remedies. In *chronic synovitis*, mild counter-irritation, such as slight blistering, may be first tried. Under this plan the fluid will frequently disappear. Should a change not take place, it may be desirable to introduce a very small trocar and canula, and evacuate the fluid. Provided the joint be kept perfectly at rest, there is no risk in this proceeding. The ceratum hydrargyri compositum, with pressure, is often very useful. No constitutional treatment seems to be required, unless the health is suffering.

In *chronic disease of the cartilage*, there is no local treatment at all to be compared to the application, under pressure, of the compound mercurial ointment.

This should be spread thickly on a piece of lint, and fixed over the whole of the joint, beneath the splint, by strapping firmly and evenly applied. It should be changed once or twice a week, and the splint re-applied. The state of the constitution should at the same time be attended to. When any obvious disturbance of the health has been remedied, cod-liver oil and iron, with Dover's powder at bed-time to insure rest, may be given. Change of air, and especially to sea air, is invaluable.

If this affection be not stayed by the means employed, it runs on to suppuration, and constitutes one of those severer forms of knee disease in which the question of amputation presents itself. My opinion is entirely adverse to the adoption of this course; and in the same condemnation I include resection. But this statement must receive a practical qualification. In the case of patients of the upper classes, to whom all the requisite care and attention can be given, I believe these operations are scarcely ever needful; but among the poor, where the means requisite for the successful conduct of the case are not attainable, it may be advisable, in practice though not in theory, to remove the limb. I say to remove the limb, because in the case of the knee joint I give the preference to amputation in the growing child. If resection be performed, I believe the limb is not ultimately more useful, since, the epiphyses being removed, it cannot grow in due proportion; and thus a more serious operation is incurred for no corresponding advantage.

The plan of treatment I adopt is one that I first saw used by Mr. Key, in 1842; namely, making free incisions into the joints to evacuate the contents. The synovial membrane having lost its peculiar characters, the same amount of disturbance does not arise from its exposure, as if it were in a healthy state. The procedure, therefore, is free from any particular risk. An incision may be made on one or both sides of the patella, and the joint treated as a large abscess. The most favourable issue, and one which may generally be looked for, is perfect anchylosis.

During this treatment, of course the child's powers should be freely supported, and if the case take an unfavourable course, and the patient threaten to sink, amputation can be resorted to at last.

2. *Diseases of the Hip.*—Among the miserable objects which make their appearance at the out-patients' room of a large hospital, none more painfully appeal to our commiseration than the children who present themselves, in such numbers, with hip disease. Our pity is increased by the feeling that, in great measure, the sufferers are beyond our aid; not solely because of the untractable nature of the malady, but chiefly from the impossibility of carrying out the necessary treatment. An hospital set apart for hip disease would scarcely suffice for the requirements of this Metropolis, so absolute and prolonged is the rest demanded for its thorough cure.

The early symptoms of affections of the hip joint are of a most insidious kind. The children are generally said to be suffering from " growing pains," for many months before the disease assumes an active character; or the child is one day found to be slightly limping, when, upon questioning it, it complains of a little pain at the knee, but of none whatever at the hip. Indeed, so characteristic is this pain at the knee, as a symptom of hip-joint disease, that sometimes, during the greater part of its course, it is the only pain complained of. Sometimes there is even swelling of the former joint, and remedies have been applied to that part, when the hip has been the seat of the mischief. When attention has been drawn to the case, there will generally be found a flattening of the nates on the affected side, owing to the wasting of the glutæi muscles; and when the child is in the erect posture, a peculiar appearance may be observed, which is very characteristic. The folding in of the skin, which marks the junction of the nates with the thigh, is on a lower plane on the diseased than on the healthy side, owing to the relaxation of the parts, and is absent towards the outer part of the limb. The child's general attitude is shown in the woodcut (Fig 1).

Fig. 1.

When the little patient stands with its face towards us, the fissure in the groin on the diseased side seems partially obliterated; and the skin of the thigh appears to be on one continuous plane with that of the abdomen. The appearance is as if a swelling within the capsular ligament of the joint threw all the parts forward, as shown in the woodcut (Fig. 2).

On taking a side view of the child (as presented in the woodcut, Fig. 3) the hip and knee joints are both observed to be bent; the foot being brought slightly forward, and the toes just touching the ground. The leg appears as if it were too long to be placed in

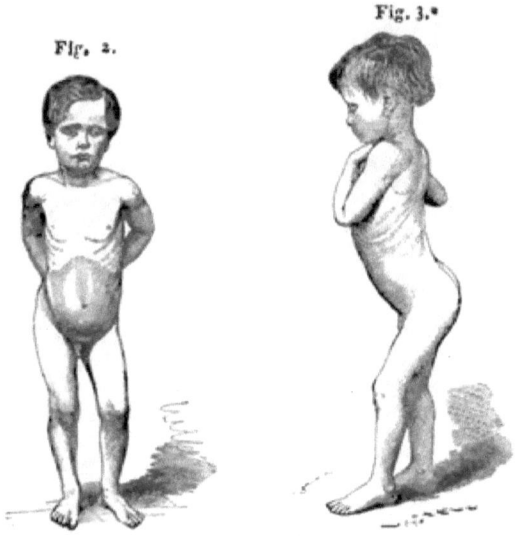

Fig. 2.

Fig. 3.*

the straight position. At the same time, the trunk is inclined a little forward.

Causing the child to steady itself by its opposite hand, the surgeon should place one thumb on the anterior superior spinous process of the ilium, and attempt to flex the femur upon the pelvis, when he will find that the whole pelvis seems to move; the pivot on which the motion takes place appearing to be

* In this figure, the knee-joint is not so much bent as it commonly is; but the woodcuts are faithful representations, taken from photographs.

the articulation of the os innominatum with the opposite femur. If the child be supported on the unsound leg, and the sound one moved in the same way, the contrast between the two is most remarkable. This symptom may be looked upon as perfectly diagnostic of hip-joint disease even in the earliest stages. It is one which I have never found to fail me. The explanation appears to be that there is a spasmodic contraction of the muscles to fix the head of the bone on the acetabulum, thereby preventing motion, which, if it occurred, would be attended with pain. Viewed in this light, the immobility of the joint appears a beautiful instance of instinctive muscular adjustment. When the child is placed under the influence of chloroform, this fixity of the joint disappears; and the diseased limb can be moved almost, or quite, as freely as the sound one. I have also found this test an unfailing criterion in simulated hip disease from hysteria in adults.

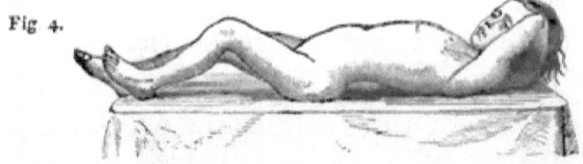

Fig. 4.

The most characteristic appearances however are presented when the child is lying on its back. And I may here observe that we cannot exercise too great caution in thoroughly examining these little patients, stripped of all their clothing, whenever the parent complains that the child is lame. The woodcut (Fig. 4) shows very clearly the position taken by the child when lying. The shoulder and buttocks rest upon the bed, causing the lumbar and dorsal vertebræ to form an arch under which the hand can be placed. The knee, also, being raised, the foot rests upon the heel, forming another arch. If the limb be straightened, which is attended with a great deal of pain, it is found to be about half an inch longer than the other. The explanation of this fact has not been satisfactorily shown; it is attributed by some to a twisting of the pelvis, but I incline to ascribe it rather to effusion within the joint, pushing the head of the bone downwards. Upon attempting to move the joint, the symptoms resulting from its immobility are even more obvious than when the child is standing. On pressing the hand deeply into the pelvis just above Poupart's ligament, there is felt a thickening, like a mass of enlarged glands, and pressure here is attended with pain. There is also pain on pressure behind or over the trochanter major; but on attempting to rotate the thigh, or on pressing the limb upwards, there is seldom any uneasiness. The whole leg appears wasted, and thinner than its fellow, and seems to have lost its "tone."

This is unhappily the condition in which the joint is found at the earliest period at which the surgeon is generally consulted. As the disease advances, the lengthening of the limb is converted into a shortening, probably from absorption of the head of the bone; there is a great increase of fulness about the joint itself; the pain grows much worse, being especially severe at night, and the child is unable to walk about. Generally, also, it begins to suffer constitutionally, but not always. When the case has arrived at this stage, suppuration soon takes place. The limb then becomes more flexed, the muscles wasted, and the region of the hip presents a distinct enlargement from the presence of matter. The abscess points, in an obscure manner, on one or other aspect of the thigh, most frequently on the fore part, in front of the tensor vaginæ muscle. Dislocation of the head of the bone follows, causing a projection upon the dorsum of the ilium, the thigh being thrown over its fellow, and the toes turned inwards. Great shortening of the limb, of course, results. Hectic sets in, and continues until the bursting of the abscess affords a temporary relief.

If the case were left to itself, and the vital powers did not succumb, the discharge after a long time would cease; ossific matter would be thrown out, the child recovering with the limb anchylosed in the bent position.

Sometimes hip disease occurs in children in an acute form. Synovitis of a severe character sets in suddenly, and speedily runs on to suppuration. I have seen death occur in this way within three weeks, no other disease being present. Generally this affection arises from a blow or other accident; but sometimes no such cause can be detected, and the nature of the disease is in these cases very obscure.

The attempt to determine the first structure attacked in hip disease has taxed the ingenuity of surgeons for many years. Every structure entering into the composition of the joint has in turn been fixed upon. The synovial membrane, the articular cartilages, the round ligament, the head of the bone, have all been supposed to be the original seat of the disease. But it seems now to be generally agreed upon that here, as in other joints, disease may originate in either of the structures named, and that hip-joint disease is not an affection of a peculiar character. Practically, the question what structures are primarily or secondarily affected is not important.

I cannot do better than introduce the following three cases, as affording a good account of the condition of the diseased joint found after death.

Case 1.—Acute Disease of Hip Joint.—W. S., aged two years, was admitted into Guy's Hospital, under Mr. Cock's care, in July, 1856. Three months before admission, he had scarlet fever, and soon afterwards appeared lame in the right leg. The hip joint was slightly contracted. Five days after admission, the boy was placed under the influence of chloroform, the leg straightened, and a splint applied. Sickness occurred soon afterwards, as is not uncommon after chloroform, and on the following day he was very ill; being in a listless, sleepy condition, only roused by shaking, and then with difficulty. These symptoms continued until two or three days before death, when complete coma set in.

On the post-mortem examination, fluid was found in the ventricles of the brain, and the lungs were tuberculous.

The synovial membrane of the hip joint was of a red colour, highly vascular, and soft. It was bathed in a purulent lymph, which also flowed from the cavity of the joint, as a thick, pink-coloured fluid. The ligamentum teres was very soft, and easily torn. At the upper part of the acetabulum, the cartilage was slightly ulcerated, producing a rough eroded surface. The parts around the joint were quite healthy.

Case 2.—Strumous Disease of Hip, Fractured Thigh, Tubercular Arachnitis, Absence of Kidney.—A. B., aged eight years, was admitted November 11th, 1854, into Guy's Hospital, under Mr. Hilton's care. He was a very sickly, emaciated child at the time of his admission; had been subject to hip disease for two years, and walked on a crutch. He was taken into the hospital on account of a fracture of the thigh on the same side, at its upper part. Sinuses existed around the joint. No repair of the fracture took place. The hip joint suppurated, and the emaciation became extreme. Latterly, symptoms of organic disease supervened, and during the last few days of his life, convulsions and coma. *Post-mortem examination.*—Body extremely wasted, abdomen contracted, right leg much shorter than the left, and displaced. The surface of the brain was covered with a thin yellow lymph, and the arachnoid had small granular deposits of lymph upon it. The pia mater was everywhere covered with tubercles, particularly in the deep sulci of the convolutions; on the left side, in the fissure of Sylvius, they formed a complete layer upon the membrane. At the base of the brain there was considerable effusion of lymph. The lungs and abdominal glands contained tubercle. Only the right kidney existed, and it was about equal in size to that of an adult.

The right leg was much shorter than the left, owing

to the head of the femur being thrown out of the acetabulum, and to the over-riding of the fractured bones. The acetabulum and a large part of the dorsum of the ilium were black and carious, and on this diseased surface lay the head of the femur; this also being black, necrosed, and much wasted. The fracture was in the upper third of the bone. The end of the upper portion projected outwards, and had become quite rounded. The end of the lower part lay to its inner side, and was loosely attached to it by fibrinous exudation and new bone.

Case 3.—Chronic Disease (Necrosis) of Hip Joint, Fatty Liver, Lardaceous Spleen.—W. W., aged nine years, was admitted into Guy's Hospital, under Mr. Hilton's care, on the 8th August, 1855. Six months before admission, the symptoms of hip disease commenced; no history was taken of the case during the year in which the lad was in the hospital; but he, for the most part, kept his bed, with a constant purulent discharge flowing from the left thigh. On the night of July 23rd, a sudden gush of blood took place from the wound; and it appeared as if hæmorrhage had been for some time going on. The bleeding was stopped, but in a short time the boy died.

Post-mortem Examination.—The body was extremely wasted, the upper part of the left thigh much enlarged, and covered with sinuses through which clots of blood were protruding. A large abscess surrounded the superior portion of the femur, passing back to the buttock, and upwards beneath the psoas muscle. The purulent matter was mixed with clots of blood, which existed in large quantity around the femur and beneath the pelvic fascia, in the suppurating cavity which had there formed, and which communicated through the acetabulum with a cavity within the pelvis. A jet of water, thrown into the common iliac artery, poured out in a full stream through the fistulous openings, showing that a large vessel must have been ulcerated. The femoral and profunda arteries, and the commencement of their branches, seemed entire; therefore it was probably the gluteal that was opened. There was much fibrous tissue, forming a solid hard mass around the joint. Enclosed in this were the muscles, which appeared completely degenerated, being partly changed into fibroid tissue and partly into fat. The bones forming the joint were completely necrosed and broken, so that the head of the femur, or the remains of it, were seen from within the pelvis, and the finger could be passed through the acetabulum into the thigh below; in fact, the abscess in the thigh, and that in the pelvis, communicated in this part. The head of the femur, the neck, and upper part of the shaft, were in a state of necrosis, and in loose pieces; as were also the acetabulum and the rim of the pubes and ilium. The epiphysis was completely detached, so that the neck of the bone rested in the socket. Lying at the side of this, were also two loose pieces of bone, forming part of the cap of the acetabulum.

Treatment of Hip-joint Disease.—In the acute form of the affection, by far the least frequent, the ordinary method of combating acute inflammation may be resorted to. If these succeed, the disease would then come under the modes of treatment appropriate to the chronic form; but in the few cases I have seen, the symptoms were so severe and rapid as to render the most active treatment utterly inefficient. The children appeared prostrated at once; suggesting the idea of a general poisoning, rather than of a merely local inflammation.

In the chronic form of the disease, it is, in the first place, absolutely necessary for the future welfare of the patient that a splint should be applied to the joint. Formerly, this was done under very disadvantageous circumstances; as the following case illustrates.

Case 1.—W. H., aged five years and a half, came under my care at the Surrey Dispensary, in Nov. 1851; a delicate-looking boy, with symptoms of disease of the left hip-joint. He had complained of pain in the left knee for twelve months; for the last six months he had been under medical treatment, but had been allowed to get about as usual. For fourteen days before I saw him he had walked very lame, and cried a great deal at night. He was attended at his parents' residence; and under these circumstances, the disease being far advanced, I despaired of attaining a satisfactory result. The knee was bent, and the thigh flexed on the pelvis; and when an attempt was made to straighten the limb, the lumbar vertebræ became very much curved. Chloroform was not administered as freely then as it is now, and I was afraid to avail myself of it. Accord-

T

ingly I applied the long splint, but was not able to bring the femur into a line with the pelvis, and for a month scarcely any improvement in its position took place. His health, too, began to suffer, and the pains at night increased. I therefore ordered him cod-liver oil twice a day, and Dover's powder at night. The splint was reapplied every fourteen days, and gradually I found that the femur could be brought into a straight line with the body. The lumbar vertebræ lost their curve, and after six months' perseverance with this treatment, the boy was able to get up and walk about free from pain, and with a straight leg.

All trouble in applying the splint is now removed by means of chloroform. When the child is placed under its influence, the limb can almost always be brought at once into a straight line with the body; and this is the plan I now universally adopt. Sometimes an aching pain is complained of for a few hours after the splint has been thus applied.

The objects to be attained by means of the splint are—

First. To insure perfect rest, thus relieving pain.

Secondly. To avert the formation of matter; or if it has formed, to determine the point at which it shall come to the surface: and

Thirdly. To secure as useful a member as possible.

The best splint for attaining these ends is either the common straight one used for fractured thigh, or a japanned tin splint made to fit the outline of the hip and thigh, which can be shortened or lengthened to suit different patients. The mode of applying these splints is the same. They should be carefully bandaged to the limb, from the foot to the lower part of the thigh, and another bandage should be applied around the pelvis. I believe this plan to be infinitely superior to the practice of enveloping the joint alone in a pasteboard or gutta-percha case. The splint should be re-adjusted every three or four weeks, and kept on for many months, the child being confined to bed. I have patients now under my care at the Infirmary, who have been lying there for months, yet in whom, if the splint be removed, pain immediately returns, and the limb begins to be slightly drawn up. The recumbent posture, though maintained for so long a time, does not injure the health. The little patients look well, eat and drink heartily, the visceral functions are perfectly performed; they lie quite flat upon their backs, and play with the toys set before them, having free use of their arms.

By thus insuring perfect quiescence of the joint, the progress of the disease is often arrested, and suppuration does not take place. If it have already occurred, the splint seems to direct its course towards the anterior part of the limb, and to insure its pointing in front of the tensor vaginæ femoris, which is by far the most convenient situation. Even if an opening have been formed, and dislocation have occurred, it is still advisable that the limb should be rendered straight, if possible, under the influence of chloroform, and the splint applied. By this means anchylosis will occur in the straight position, by far the most desirable as regards the future usefulness of the member. By some it may be doubted whether this be the case, as the patient is unable to sit down comfortably; but the inconvenience in this respect is not great, and the advantage in the erect posture far outweighs it.

In addition to the use of the splint, other external applications may be had recourse to if necessary. Leeches have been recommended if there be pain, and I was formerly in the habit of making use of them. Lately, however, they have not seemed to me to be necessary. If abscess be forming, poultices can be applied.

Internally the exhibition of bark or iron is useful. Dover's powder may be given at night to relieve pain, and children of four or five years old will often take five grains for a dose without any ill effects. This internal treatment is applicable to all stages of the disease, as also are generous diet and sea-air; but the latter should never be suffered to take the place, as parents often wish, of the unremitting application of the splint.

When an abscess has formed, and is approaching the surface, the question occurs: Should it be opened? The plan recommended for the treatment of psoas abscess may be adopted, and the pus evacuated by means of a valvular opening. Upon the whole, however, it is doubtful whether, if the patient's health be not suffering, it is not better to allow the abscess to discharge itself, which it will do at proper time and place. In the more advanced stages of the disease, when sinuses have formed, and bone is to be felt through some or all of them, a question may occur whether the head of the bone should be excised, particularly if dislocation have taken place. If the excision be designed simply for the removal of an extraneous body which is keeping up irritation, there can be no doubt of its advisability, as by this means we give the patient the best chance of recovery. But if it is to be performed as the excision of a joint, it would be necessary, also, to remove all other diseased portions of bone; and when we reflect how constantly the acetabulum is involved, how frequently this portion of bone becomes perforated, and the disease extends into the pelvis, we cannot be surprised that the proposal is looked upon discouragingly by surgeons generally; or at all events, that the operation is but sparingly performed. Few cases are more promising in theory, but in practice few are more difficult of performance, or more unsatisfactory in their results. I have witnessed an attempt to remove the head of the bone from the dorsum ilii, when it had never quitted the acetabulum.

The following cases, taken indiscriminately from my note-book, may serve to illustrate the advantages to be obtained by the use of the straight splint.

Case 2.—G. W., aged nine years, came under my care in June, 1851, and was attended at home. He was tall for his age, and pale, but before his present disease, was stout and healthy. Six months previously, he complained of pain in his left hip after running; and there had been slight swelling of this joint at times. He had also complained of what were termed "growing pains." The usual symptoms of hip disease were present; and he suffered great pain at night. The straight splint was immediately applied, and notwithstanding the pain occasioned by the straightening of the limb (without chloroform), instant relief was felt as soon as the manipulations were concluded. He continued in this position for two months, when a sharp attack of pleurisy supervened, also on the left side, terminating in empyema, which burst externally below the nipple. He continued to wear the splint for six months longer, and was then quite well, with scarcely any stiffness of the joint. I saw this boy five years afterwards, and he was able to use the limb almost perfectly.

Case 3.—T. D., aged three years and a half, was attended at his own home, coming under my care in June, 1851. He was a delicate-looking boy, and had been the subject of strumous ophthalmia. The history obtained from the mother was, that a year ago he complained of pain in his left leg when walking, and especially about the knee. During the last three weeks, he has had so much pain on any movement of the thigh, that he has kept his bed. The alteration in the shape of the joint was very slight; he complained of a good deal of pain at night, and was very restless. The thigh was flexed upon the pelvis. The straight splint was immediately applied, and Dover's powder was given every night. The splint was continued until the end of April in the following year, a period of ten months, when my report states that he walks as well as any one could with a stiff hip. There was neither pain nor swelling, and he was well in health.

Case 4.—E. D., aged five years, came under my care in November, 1851; being treated at her own home. She was a hearty-looking child, and always had good health. Seven months ago, she complained of pain in her knee, which swelled. Blisters, &c., had been applied to it. She had only obscure symptoms of hip-joint disease; very slight deformity, and scarcely any pain, and she could bear the whole weight of her body on the diseased leg. But, on attempting to flex the femur, the pelvis moved with it. The straight splint was applied, and cod-liver oil given. In a month's time, the child began to have rigors, and a fulness became perceptible in front of the tensor vaginæ femoris. This swelling increased, and the pain became severe. The splint, however, was continued, and Dover's powder was given at night. On the 24th of March, the abscess burst. There was at that time an inch shortening of the leg, and the child always slept with the toes of the diseased limb supported by the opposite foot. Not until thirteen months afterwards did I discontinue the splint, when the child could walk without discomfort, the hip being anchylosed in the straight position.

3. *Disease of the Elbow Joint.*—The diseases of the elbow, in children, are strictly analogous to those of the knee. Synovitis, and its consequences, ulceration of the cartilage and disease of the articular ends of the bone, are the prevailing

affections. It is unnecessary, therefore, to enter on a detailed description of them.

The swelling produced by effusion of fluid takes the form of the capsular ligament, and extends upwards beneath the triceps muscle, on the posterior aspect of the humerus. There may be detected, especially in the earlier stages of the disease, a tender spot just in front of the internal condyle, which is a useful guide in the diagnosis. Disease of the elbow is much rarer than affections of the knee or hip, and practically, also, is much more amenable to treatment.

The child naturally places its arm in a slightly flexed position, but not by any means in the most useful one for after life, should anchylosis occur. Accordingly, the first and most important part of the treatment is to place the elbow at a right angle, by means of any splint that may be preferred. An angular one, fitted with a screw, applied on the inner side of the arm, or a piece of leather, cut at right angles, and placed for the arm to rest upon, are among the most convenient. At the hospital, I am in the habit of using millboard cut in this way, and find it answer every purpose.

The limb being thus placed at rest, and prepared for the worst, we may adopt any medication that seems to be indicated. The principles which should direct the treatment are the same that are laid down for disease of the knee. Should the disease progress unfavourably, and the joint threaten to become permanently disorganized, change of air should be tried; and the sea-side is to be recommended. I believe there are few cases in which this course, unless too long deferred, will not save the joint; but if it be not practicable, or fails, operative interference becomes necessary. Amputation of the arm, in these cases, is out of the question; the remedy is excision of the joint. This operation is almost invariably successful, and gives the child a very useful member. It is astonishing, indeed, how large an amount of motion may be secured by it, and the ultimate issue of the worst cases is thus often better than that of the less severe. In the last instance in which I excised the joint, and in which a large portion of the ends of the bones was removed, the amount of motion obtained was very great.

Resection of the Elbow Joint.—The child being placed in a convenient position, lying partly inclined upon its face, so as to present the back part of the joint to the surgeon, the first incision may be made in one of several ways, each of which has its advantages in different cases. When a large space is required, the H or the T incisions may be used, but generally, the single straight incision will be found the best. This may be from four to six inches in length, the olecranon forming the central point. The incision should be carried down to the bones in its whole length, and the tissues separated from them. The whole of the diseased portions of the bones are then to be removed, either by Mr. Butcher's saw, or by bone forceps. Any bleeding vessels are to be secured, and the divided soft parts brought together by suture. The arm is to be placed in a splint, bent at a right angle.

The subsequent treatment of these cases demands great care and attention; the surgeon is frequently annoyed, after the lapse of several weeks, when the cure seems complete, to find a sinus existing, and pieces of diseased bone requiring removal.

III. CONTRACTED JOINTS.

Children are frequently brought to us with contracted joints, the result of previous disease, all inflammatory symptoms having subsided. The cause of the parents seeking advice is the inconvenience of the position of the limb. In these cases a very serious responsibility rests upon the surgeon: on the one hand, it is often the case that a great benefit might result from altering the angle at which the joint is fixed; on the other, serious danger may arise from breaking down the new tissue. The advantage is considerable; but the risk is great. Unfortunately the elements on which a judgment might be formed are very difficult of access. The union is sometimes bony, sometimes fibrous. Even in the latter class of cases interference is not always safe; in the former it is of course out of the question. But even this preliminary point cannot be decided with certainty. A very minute and accurate history may assist us somewhat; for instance, the amount of inflammation that accompanied the original disease, and the length of time during which it lasted, are very important. Also, if an abscess have formed and burst, the case is much less favourable; but none of these circumstances can give us assurance of the condition of the joint. A short and slight attack of inflammation without suppuration, may leave it irremediably fixed; a tedious suppuration may result in a merely fibrous union that may be safely broken down. The decision must rest ultimately upon the surgeon's individual experience and tact. In every case gentle extension may be tried experimentally, to see if any symptoms of yielding appear. If even a slight movement can be thus detected, it may fairly be presumed that the union is ligamentous only, and under such circumstances further proceedings may be confidently undertaken. But even in this case I believe that a separation of the distal bone from its epiphysis sometimes takes place. If this occur in the knee, the tibia slips partly back, projecting into the popliteal space, and the little patient recovers with a limb straighter, indeed, and more useful than before, but more deformed.

These cases are likely to become more and more rare as the principles of surgical practice become better understood; since it is, of course, only through want of treatment, or bad treatment, during the original disease that such a condition can arise. Properly treated, a diseased joint is kept throughout in the best position in which it can be placed for future usefulness, if anchylosis should occur.

The method of dealing with these cases varies according to the joint. The elbow requires to be flexed to a right angle, if it is straight; the hip or knee, if bent, to be straightened, the child of course being under the influence of chloroform. In the case of the elbow, the patient's arm being held firmly by an assistant, the surgeon should grasp the humerus with one hand, and the forearm with the other, and bend the elbow to its full extent. Some recommend that this should be done forcibly and rapidly; but I prefer to do it gradually. The arm should then be brought to the position in which it is intended to remain, and a bent splint, well padded, applied for ten days or a fortnight. Some surgeons affirm that a useful degree of motion may be acquired by removing the splint, and mechanically moving the joint to and fro for a short time every day, as soon as the symptoms of inflammation have subsided. The extreme pain of this treatment has always seemed to me to be a sufficient reason against its employment in the case of children of tender years. Nor is there any satisfactory evidence of its efficiency.

If the hip be fixed in a bent position, the child should be placed at the edge of a bed, and the pelvis being firmly held by an assistant, the femur should be gently extended to its utmost limit. As this is done, the surgeon will feel the sensation of fibrous tissue being torn, which once felt can never be mistaken. The thigh should then be placed in a line with the trunk, and a straight splint applied for three weeks or a month. During this period it may be removed two or three times, and friction used to the joint, which gives great comfort.

The knee should be treated in the same way; but in the case of this joint there arises the question whether the tendons of the hamstring muscles should also be divided. When they appear remarkably tense, I think it is better to free them by a tenotomy knife; this being preferable to tearing their attachments, or lacerating the muscles, one or other of which must otherwise occur. If the tibia should unfortunately slip backwards, a very clever mechanical contrivance has been invented by Mr. Heather Bigg, by means of which the weight of the trunk is thrown directly on the tibia, and any inconvenience from the malposition of the bones averted.

The amount of force which may be employed in straightening joints is very great. The surgeon need not hesitate, when he has reason to believe the case is a suitable one, to put forth his whole strength. Some of these cases, however, even when the anchylosis is not osseous, defy all attempts to produce extension. Whether, as the patient grows older, some alteration may not take place in the joint, which may render the operation more feasible, I do not know. This method of treatment has not yet been practised long enough to enable us to speak on these points with confidence.

CHAPTER XVII.

DISEASES OF THE SKIN.—INFANTILE SYPHILIS.

THERE are a certain number of skin diseases more frequently met with in children than in adults, without a reference to which this work would be incomplete. At the same time, as affections of the skin are treated of at length in so many books specially devoted to the subject, it is not necessary for me to describe them systematically. I shall not, therefore, enter upon a classification of the various forms which these diseases present: the great point in respect to diagnosis is to distinguish those which are due to the presence of epiphytes, and are therefore local in their character, from those which are constitutional. Of the former class there are four which are met with in children: they are the affections called porrigo; namely, porrigo lupinosa, decalvans, scutulata, and favosa. In each of these diseases a peculiar fungus may be detected by examination under the microscope, probably a modification of the same vegetable parasite. These, with scabies, which is the result of the presence of an ectozoon, constitute the important class of the truly local skin diseases. And of them I shall first speak.

Although in respect to their immediate cause these

diseases are local, yet it is to be observed that they are generally met with in depraved states of the constitution, since it is in such states that the fungus finds the most suitable nidus. They are, however, communicable to healthy individuals, if cleanliness be not observed, and now and then cases occur that defy all explanation. When once the fungus has established itself, there is no class of diseases which are more difficult to treat. Of course local treatment is essential; but combined with it constitutional measures are generally indicated.

The principle of treatment is to apply either some caustic substance, which, though not sufficiently powerful to injure the skin, will destroy the vitality of the fungus, or some oily matter to arrest the processes on which its existence depends. Both these plans are equally effectual. Whichever is used, when the affection is on the scalp, the skull should be fitted with an oil-skin cap. Of caustic substances the best is sulphurous acid, an ounce to six ounces of water, or stronger if required; or a solution of the sulphate of copper four grains to the ounce, may be employed. These should be applied by being poured into the mother's or nurse's hand, and well rubbed in for ten minutes or a quarter of an hour, over the whole space involved in the disease. This is to be done every night, and the oil-skin cap applied afterwards. The duration of the treatment varies immensely. It need hardly be said that before the use of any other applications, a fair trial should be given to soap and water. I always employ the best brown Windsor soap, and it is astonishing how frequently this suffices for the cure, and no other treatment is found necessary.

Oily substances may be used instead of the caustics; but they are far less cleanly. Any kind of oil will do; cod-liver oil I think is one of the best, but they are all practically objectionable. The unguentum hydrargyri nitratis answers very well, and the unguentum æruginis combines both plans of treatment. The formula for it is—

    ℞ Æruginis . . . . . ℨj
       Adipis . . . . . . ℨj Ft. Unguentum.

When constitutional treatment seems to be indicated by disorder of the chylopoietic viscera, the following powder may be given every night or every alternate night:—

    ℞ Sodæ carb. exsiccatæ . . . gr. iij.
       Hydrargyri chloridi . . . gr. iss.
       Pulveris cretæ co. . . . . gr. v., Misce, ft. Pulvis.

For scabies may be used, instead of the ordinary ointment, a lotion formed by boiling lime and sulphur together in water.' After the child is well washed in the evening, this lotion is to be rubbed in before a fire for twenty minutes or half an hour, and the child then well washed again. One such application suffices for a cure. The following is the formula for this invaluable lotion:—

    ℞ Quick lime . . . . . . ℥j.
       Sublimed sulphur . . . ℥ij.
       Water . . . . ℥x., Misce.

Boil, and stir frequently with a wooden spatula until the two former ingredients have perfectly combined. Decant the fluid, and keep in a stoppered bottle.

The second class of skin diseases embraces all those in which no epiphytic growths have at present been detected. It is not unlikely that many of this class may yet be transferred to the former; but until evidence is obtained of the presence of a fungus, they must be practically regarded as of a constitutional character. The most common of them in children are eczema, impetigo, and herpes; psoriasis and lepra are occasionally met with. The general principles of treatment in all these affections are the same, but the results are very variable. Sometimes a speedy cure ensues, at others the disease defies all remedies. Local treatment is not to be relied on, though in hospital practice I have sometimes found the ung. picis liquidæ very useful in lepra and psoriasis. To relieve itching, if demulcents, such as linseed-tea, fail, a lotion composed of an ounce of glycerine and a drachm of bicarbonate of soda to a pint of water may be applied.

The utmost cleanliness must be enforced; warm baths two or three times a week are very useful. A large number of cases recover under these means, with attention to diet and to the state of the digestive organs. In respect to diet, a free use of fresh vegetables and ripe or cooked fruits is of great importance. In the more troublesome forms of eczema, small doses of arsenic seem to be the most effectual remedy;—a quarter of a minim of Fowler's solution twice a day for a child twelve months old, is a safe dose, and seems to be attended with very beneficial results. The practitioner will, of course, be on the watch for any of the specific effects of the remedy, and guard against its accumulation in the system. Full details respecting this class of cases, however, are to be found in the works of the dermatologists.

INFANTILE SYPHILIS.

Terrible as are the ravages of syphilis in the adult, it is an even more pitiable and fearful sight to witness the sufferings of an infant labouring under this inherited curse. Sometimes children are born with evident symptoms of the disease; miserable, thin, wan, cachectic-looking objects, covered all over with a copper-coloured eruption, with the epidermis scaling, with a continual discharge from the nose, often a fissure at the angle of the mouth, and condylomata about the anus—these children seem only brought into the world to die. They are, for the most part, born prematurely; in which case, they almost always die immediately. This, however, is not the usual form in which infantile syphilis is presented to the surgeon. On the contrary, the children are frequently, at the time of birth, healthy, hearty little things. In about a fortnight, or from that to two months, the first symptoms make their appearance. The child is supposed to have caught cold; it begins to snuffle, and this continues for some few days before any other symptom appears. This coryza, caused by swelling of the mucous membrane of the nose, prevents the child from breathing during the act of sucking; it also begins to get fretful, and loses flesh. Shining patches next present themselves on some part of the body, especially the cheek and forehead, or about the anus and scrotum; but in many cases they appear simultaneously all over the body. These are of a coppery colour, and soon become scaly, like psoriasis. Fissures form at the angles of the mouth, and about the anus; in fact, at each extremity of the intestinal canal. These fissures frequently bleed. The child now looks feeble and cachectic, and the voice, as indicated by the crying, is shrill and piercing. This peculiar sound is most characteristic of the disease. There is a copious secretion from the meibomian glands; the eyelids stick together, and fissures form upon them. The features have a peculiar old-man appearance. Frequently flat tubercles, which appear to be ulcerating condylomata, form about the anus. Sometimes these condylomata, with fissures about the anus, are the only manifest symptoms of constitutional syphilis in a child. The eruption may be either a simple stain of the skin (maculæ), or scaly, or pustular. The latter is very rare, and is almost always indicative of a fatal issue. When pustules are present, they seem to affect chiefly the palms of the hands and soles of the feet. There is sometimes, as in adults, an ulceration between the toes or fingers. Dr. West states that he has seen the palate bone diseased. I have never met with any case in children where syphilis has advanced so far as this. But in describing the diseases of the mouth, reference has been made to a form of ulceration of the palate, in which the bone becomes implicated, but which does not appear to be syphilitic.

In forms like that above described, hereditary syphilis is universally recognised. But it may be questioned whether many other forms of disease might not truly be ascribed to the same cause. Very frequently it happens that the children of fathers who have had syphilis, are subject to obstinate skin affections, principally of the scaly character; these will sometimes defy all treatment, even the thorough employment of mercury. Indeed, it is impossible to say where the effects of the wrong thus inflicted, on his offspring by the parent, end, or when, or in what form, the virus may break into activity.

Mr. Hutchinson, in the *Pathological Transactions* for 1858 and 1859, has described various effects of syphilis upon the temporary and permanent teeth, especially the latter. I need only refer my readers to his monographs on this subject. My own observation leads me to believe that his views have some foundation, and he has at least collected facts which tend to confirm the opinion that many obscure diseases met with during youth may have a syphilitic origin.

The *Diagnosis* of hereditary syphilis is generally very easy. Now and then, however, a case occurs, which if seen only once, may present a good deal of difficulty. The affection of the skin sometimes resembles psoriasis, and when the child is plump and healthy, with only a slight degree of eruption, it alone would be an insufficient guide. But the symptoms taken altogether are very characteristic, especially the snuffling and the soreness about the anus. The skin, also, has a peculiar waxy hue, which is seldom absent.

The *Prognosis* of these cases is very favourable; though fatal results sometimes occur. Among the poor the risk is increased by the difficulty of carrying out the necessary treatment. The death, in these cases, is attributed by my friend, Dr. Wilks, to a specific affection of the internal organs, chiefly of the liver and lungs, in which a peculiar deposit is found.

In respect to the *Treatment*, nothing can be more certain than that there is a specific; and nothing more satisfactory than its effects. I invariably use, and have done for some years, the mercurial ointment spread on lint and bound over the abdomen. This is to be changed every second or third day, for the sake of cleanliness; and it should be continued until a fortnight after every symptom of syphilis has disappeared, which may occupy two or even three months. This seems to be by far the best plan of treatment. I have never found it necessary to give mercurials to the mother, or the fashionable grey powder to the child. This latter remedy, though it often acts admirably, will now and then purge, which interferes with its administration; nor does it seem equally effectual with the ointment. The fissures about the mouth, and the condylomata, when present, require the local application of nitrate of silver.

Under the above treatment, the child loses all the symptoms of syphilis, and becomes fat and well, and it is seldom necessary to give small doses of iodide of potassium, or any other "anti-syphilitic."

DISEASES OF THE NAILS.

Children are liable to a peculiar affection of the nails, which one sees now and then, and which is characterized by an unhealthy ulceration around the root. It begins with a redness and swelling about the matrix, and a little oozing of discharge takes place from beneath the fold of the skin. The nail grows thin, soon begins to break off, and becomes irregular at its free extremity; frequently half its length may thus be lost, and the exposed surface begins to ulcerate. This, in an adult, would be called malignant onychia; and if by that term is to be understood an immense difficulty in treatment, there can be no doubt of its appropriateness to the disease now in question. Sometimes the surface will slough, at others it will heal, and thus it will go on for weeks and months, in spite of the most varied treatment. I have seen it in little boys of a healthy appearance, who were willing to submit to any measures to get rid of their trouble, and have accordingly tried many plans, including entire removal of the nail. This is the most efficacious for the time, but as it grows again, the disease returns. The application of strong nitric acid to the surface is of little avail: nor does the disease improve under the absence of irritants, and the use of simple water dressing. All the cases I have seen finally recovered, but the cure could hardly be attributed to any of the means employed.

When it is necessary to remove a nail, one blade of a pair of dressing forceps should be pushed down between the skin and the nail, to its root, and with a twist it may be forcibly extracted. This is easily done, and is very much less painful than might be imagined.

WARTS AND CORNS.

The surgeon is now and then consulted respecting these formations in children. The treatment is to destroy them, as in the adult. The best applications are, for warts, the nitric acid; a glass brush being used to prevent the spread of the acid to the surrounding skin; and for soft corns, the strong acetic acid.

CHAPTER XVIII.

DISEASES OF THE EAR.

OF the causes of deafness in children it is not my intention to speak at length. When not congenital, it is, moreover, rarely met with, unless accompanied or preceded by some affection of the meatus externus, or the throat. Of the throat I have already spoken. The diseases of the ear, therefore, to which it is my intention now to refer, are those of the meatus, including the cases in which the membrana tympani is perforated, and the tympanum exposed.

I. FOREIGN BODIES IN THE MEATUS.

Children seem to have a peculiar pleasure in depositing foreign substances in the ear, among the other cavities of the body. Accordingly, the most various substances are met with in this tube—peas, slate-pencil, shells, &c.; cotton wool, part of which only has been removed, or which has been forgotten, is not unfrequently seen. These substances may give rise immediately to acute symptoms, or they may remain a long time in the meatus without attracting attention. In the latter case, however, difficulty of hearing, or uneasiness in the ear, at last induces the mother to seek advice.

It cannot be too strenuously insisted on that a careful examination of the ear should in every case be made by the surgeon; without it, any relief which may be given to the child is at best merely fortuitous. The method of examining the ear is very simple. I am in the habit of employing a silver speculum, with the small end prolonged for about half an inch, such as is commonly in use; this being introduced into the meatus, by the aid either of sunlight or of the lamp depicted at p. 15, the whole of this passage and of the membrana tympani may be explored. In by far the larger number of cases, however, a foreign body, if present, may be seen without these aids; but its absence cannot be affirmed without a complete exploration by their means. This is of importance, because now and then mothers will bring their children to be relieved of a foreign body in the ear when none is present.

The *Treatment* is of the simplest kind: we trust to hydrostatic pressure to accomplish an object which the fingers and instruments of the surgeon are ill-fitted to attain. The more the meatus is pulled about by instruments, the less likely is the removal of the foreign body to be effected. Syringing with warm water is therefore the practice to be adopted. When the body has been thus dislodged, the end of a curette may be sometimes useful to remove it from the orifice; but even then any attempts which give pain to the patient should be at once discontinued. Repeated syringing may be required, as no force should be used, even if unsuccessful in the first or second attempts at extraction. If the foreign body cannot be removed by syringing, which, unless there has been previous injudicious interference, is seldom the case, it should be left to ulcerate its way out.

Any inflammatory symptoms that may ensue from the presence of the body, or the attempts at extraction, should be combated by the usual antiphlogistic treatment.

Cerumen seldom accumulates in the ears of children, though it is now and then met with, causing slight deafness. A careful examination with the speculum is required to detect its presence, though the symptoms of intermittent deafness, and slight singing in the ears, without any previous history of ear disease, might suggest the probability of its existence. It is to be removed by the syringe.

II. DISCHARGES FROM THE EAR.

These are symptomatic either of disease of the meatus or tympanum, or of the growth of a polypus.

a. Disease of the Meatus and Tympanum.—A discharge from the meatus, which in the public mind stands for an essential disease, is, in fact, merely a symptom of various morbid conditions of the meatus, membrana tympani, or cavity of the tympanum. It is most frequently a sequence of scarlet fever or measles, though it may result from cold, a blow, or irritating liquids incautiously dropped into the ear. The ordinary symptoms of inflammation of the meatus are pain, swelling, and desquamation in that canal, followed by a thin yellowish discharge. This affection is commonly known as ear-ache. The pain shoots about the temple and down the neck, and also affects the auricle, which is frequently swollen. Mastication is painful from the movement of the jaw. Upon attempting to look into the ear, the canal is generally found so swollen that the introduction of the speculum is physically impossible. The constitutional disturbance is sometimes very great. There is at this stage a thin serous discharge. If local depletion, by leeches, be employed, this swelling rapidly subsides; but under any circumstances it disappears for the most part in the course of a few days. It is seldom, however, succeeded by complete recovery; the meatus continues slightly tumid, covered with a layer of desquamating epithelium, and secreting sometimes a profuse discharge. The anatomical structure of the parts may account for the difficulty of perfect repair; the membranous meatus being a thin layer of tissue with bone immediately subjacent, like the skin over the fore part of the tibia.

When a speculum can be introduced, the membrana tympani will be discovered with one or two large vessels meandering over it. Sometimes it is covered with a scale of epithelium, hiding this appearance. The whole internal surface of the meatus, beneath the scaly epithelium, is red, shining, and bleeds if touched. If in this condition injudicious treatment, such as stimulating applications, be adopted, or if the child be in bad health, or pulled down from the effects of scarlatina or measles, one day upon looking into the meatus, a small hole may be discovered in the membrana tympani, thus indicating an extension of the disease into the tympanum. Under these circumstances exfoliation of the ossicula may take place, or if the case issue favourably, the opening may entirely close. Even should it continue open, provided the discharge ceases and the membrane recovers its healthy condition, the hearing may be very little impaired.

Before the surgeon gives a diagnosis in these cases, it is imperatively necessary that he should be practically acquainted with the appearance of the parts, both in their healthy and their diseased states. And, inasmuch as there is invariably a collection of purulent fluid in the meatus, the surgeon should always take the trouble to wash it out with a gentle stream of warm water. Frequently, by this means, he will discover a pellet of cotton wool impacted in the ear, which may very likely be the entire cause of the continuance of the discharge. Should the membrana tympani be destroyed, as it frequently is, before the surgeon sees the child, he may diagnose the nature of the case by recognising the red mucous membrane of the inner wall of the tympanum. Almost invariably in these cases, a portion of the upper part of the membrana tympani, to which the head, and more or less of the handle of the malleus are attached, remains in its position. Sometimes this portion of membrane presents a distinct lunated edge; sometimes, however, it is fallen in, and adheres to the inner wall of the tympanum. In these cases, when the patient blows, with his mouth and nose closed, air will sometimes be passed through the meatus, giving positive proof of the perforation of the membrane. But the absence of this sign is no evidence of the absence of perforation, since closure of the Eustachian tube, from thickening of its lining membrane, is a very frequent accompaniment of this condition of the ear. The small *pulsating* surface seen at the bottom of the meatus, first specially alluded to by Mr. Wilde, is also very characteristic of destruction of the membrana tympani.

In respect to the *Treatment* of these diseases, it is in all cases necessary that the utmost attention should be paid to cleanliness. Much of their characteristic obstinacy depends simply upon the accumulation, within the meatus, of the irritating discharge. This of course undergoes decomposition (which partly accounts for its offensive smell), and its effect on the internal parts may be surmised from the irritation, and even excoriation, which it so often excites at the orifice of the canal. To prevent the escape of the discharge, the child's attendants often adopt the filthy practice of stuffing the ear with cotton wool, than which nothing can be more injurious. By this means, the meatus is converted into an abscess, and the discharge pressing upon the membrana tympani, hastens, if it does not produce, its rupture.

It would be well for the patients, if surgeons who do not know the appearance of the deeper-seated parts of the ear in these various morbid conditions, would abstain from doing more than keeping the parts clean and free from cotton wool, which, indeed, is frequently all that is required. Attention to the constitutional condition is of the first importance; next, the application of local remedies. In the acute stage, local depletion by one or two leeches, applied to the edge of the meatus, affords immense relief to the patient. This may be followed by hot steaming poultices or fomentations. The meatus should be temporarily closed with cotton wool, to prevent the leech from penetrating it, or the blood from flowing into it. The milder antiphlogistics also should be given, but mercurials and antimonials are seldom needed before the age of puberty.

When the disease has passed into the chronic stage, and there is a more or less constant and offensive discharge, the first thing to do is to ascertain whether the membrana tympani is ruptured or not. If it be not, the prognosis is very favourable, and the discharge will most likely rapidly disappear. Local applications, by means of a camel's-hair brush, of a solution of the nitrate of silver (two grains to the ounce), should be made by the surgeon twice or three times a week; and an injection of weak lead lotion should be used by the nurse on alternate days. It is necessary to instruct the nurse in the use of the syringe; not to use too much force, or to use too much or too little of the lotion. A common pewter syringe, holding about an ounce, is practically all that is necessary; but, for the richer classes, an india-rubber bottle, of about the same capacity, or a gutta-percha syringe, is preferable. If the case be tedious, the lotions may be varied by alum and the preparations of zinc; or they may be discontinued for a time, and then resumed; cleanliness, of course, being unintermittingly maintained.

The constitutional treatment consists in improving the general health, even though the child may appear pretty well. For this purpose, I know of no remedy equal to the syrup of the iodide of iron, alternated with quinine. By this treatment, indeed, without any local measures besides keeping the ear clean, I have frequently arrested a troublesome discharge. If the disease defy the above treatment, there are no means so efficacious for improving the health as change of air, especially, in the autumn, the sea air.

If the membrana tympani be ruptured, the constitutional treatment is the same, though it is frequently found, that in these cases, in common with other affections of the mucous membranes, the chlorate of potash has a peculiar effect, quickly altering the action and checking the secretion. The local remedies must be used with more caution, and syringing very gently performed. Having to deal with a mucous membrane here, we can easily understand—and who has not by painful experience found it out?—that the discharge from the surface is most intractable. It may defy all remedies for months and years, till at last the patient's friends are wearied, and give up any further attempts at cure. The first thing to be done is most scrupulously to attend to cleanliness, more so than in any other cases. Changes may be rung on the various injections, but they should not be applied too strong. If they are gently stimulating, it is sufficient. Chloride of zinc (one grain to the ounce) is useful; or better still is, glycerine an ounce, and liquor plumbi a drachm, to six ounces of water. I am in the habit of frequently applying the lotions myself, besides directing their use at home, to insure their thoroughly reaching the parts. I have sometimes applied the stick nitrate of silver, fused to the end of a coiled spring, represented in the accompanying woodcut, and have found great advantage from it.

The Artificial Membrana Tympani.—When the discharge has been subdued, and the patient remains deaf, great benefit may accrue from the application of the ingenious apparatus suggested under this title. The admirable work lately published by Mr. Toynbee will afford my readers the best and fullest information on the subject. It is the less necessary for me to go into details respecting it, since it is only when the children are advancing towards puberty, that it is, in my opinion, advisable to use it.

b. Polypus.—As I have before observed, discharge from the meatus is often the only obvious symptom of the presence of a polypus; which is another very cogent reason for examining the ear in every such case. Often and often cases have been brought to me, in which ocular examination, even without the speculum, after the ear had been syringed, would have at once revealed the nature of the disease.

These polypi are either soft and friable, or hard and fibrous; in the former case, they pour out an abundant secretion, in the latter but little. The polypi appear like an exuberant granulation from diseased bone. Sometimes they are so large as to fill the whole meatus. Their appearance, indeed, has occasionally led me, at first sight, to suppose that the bone was diseased; and until a careful examination was made with a probe, which can be generally passed all round them, I have not been able accurately to determine their character. There sometimes occurs a prominence which seems to be a strumous thickening of the membrane lining the meatus, analogous to that which is met with in the mucous membrane of the nose.

Treatment.—The soft polypi will occasionally drop off of themselves, though this is not a result which the surgeon can expect. The only satisfactory means of getting rid of them is to remove them, which may be done either by the snare, introduced by Mr. Wilde of Dublin, shown at p. 20, or by the forceps. Mr. Toynbee describes a very ingenious forceps, invented by himself, which he calls the "lever ring forceps." In removing a polypus with the snare, which is the means I almost always employ, it is necessary first to ascertain, by means of a probe, from what spot the polypus grows. Sometimes a little troublesome hæmorrhage follows its removal, but this quickly subsides. However removed, polypi are very liable to return. To guard against this it is desirable to employ an injection of sulphate of zinc after their removal, with any constitutional treatment that may be indicated. For the thickened condition of the meatus, I make use of the same remedies as when the mucous membrane of the nose is affected; viz. the local application of nitrate of silver, and the administration of cod-liver oil and iron.

GENERAL REMARKS.

The surgeon should be on his guard against *Malignant Disease of the Ear*, which now and then, though rarely, attacks young children, and might, in its early stages, be mistaken for polypus; in which case, a very injurious interference might be the result.

Amongst the accidents to which the ear is liable in children must also be enumerated *Laceration of the Membrana Tympani* from boxing the ears. When this occurs, a slight discharge of blood appears at the orifice of the meatus. The rent may be seen by means of a speculum, and it should be treated simply by rest. I have never seen any ill result from this injury, and the rent heals in a few days.

The membrana tympani, also, though extremely sensitive to the touch, is sometimes injured by putting pins or other pointed instruments into the meatus, which should not be forgotten.

A frequent cause of disease of the ear, I believe, is the practice of not thoroughly drying children's heads after they are washed. This is more particularly to be attended to when there already exists any discharge from the ear. The washing, when necessary, should be done in front of a good fire at night, and the hair rubbed perfectly dry. After bathing, or the ordinary morning wash, also, especially when the health is not strong, the greatest care should be taken in thoroughly drying the hair. If water lodge in the ear, producing a very unpleasant buzzing noise, instant relief may be given by introducing into the meatus a fragment of blotting paper rolled into a coil, which absorbs the fluid.

There is another point of great importance, which the surgeon should never forget, and in respect to which his counsel is of the utmost value. I allude to cases in which children, from being slightly deaf, are thought to be stupid or obstinate. Very sad it is to think how often a child is thus punished for his misfortunes, and, it may be, irremediable injuries inflicted on the mind or temper of this poor victim of unintentional injustice. It is hardly necessary to insist upon the care which is requisite in examining the state of the hearing power in a child, or to refer to the fact that they will often say, and doubtless think, they hear a watch when they do not.

CHAPTER XIX.

CONGENITAL DEFORMITIES AND MALFORMATIONS.

THIS subject embraces so large a sphere, that an attempt even to mention all the varieties of deformity to which the human species are liable, would be an almost endless task, and one that would serve no practical purpose. I have already alluded to some of them; hare lip, imperforate vagina and hypospadias; and as this work aims at being of a practical character, it is my intention to refer only to the other comparatively common ones which may be met with in daily practice. Many of these deformities, moreover, are quite irremediable, and an account of them therefore would be of simply anatomical interest.

Commencing with the nervous system, I shall first allude to the two similar affections, encephalocele and spina bifida.

I. ENCEPHALOCELE.

The ordinary form and position of this swelling are represented in the accompanying woodcut, which was taken from a child admitted into Guy's Hospital in October, 1858. He was two months old. The pouch was transparent, and appeared to consist only of the membranes, and would therefore by some be called a meningocele. It contained fluid. The sac was then beginning to slough; subsequently it burst, and the fluid escaped, the child remaining perfectly well. The integuments began to shrink, and when he left the hospital there appeared a good chance in favour of a perfect cure. These swellings are almost always

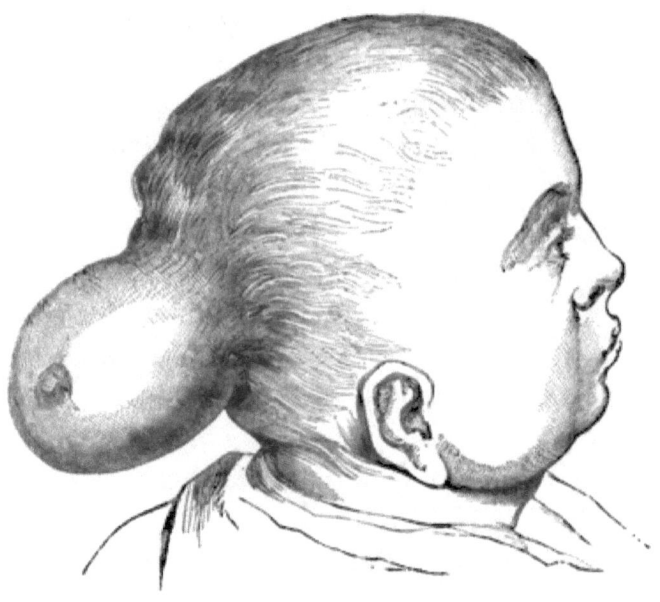

situated in the mesian line of the head, generally at the back part. They have a constricted base, but may attain any magnitude, from that of an egg to double the size of the child's head. They present many varieties, and reports every now and then appear in the periodicals of "extraordinary cases." The affection is the protrusion, sometimes of an accumulation of fluid in the meninges of the brain, sometimes of the brain tissue itself; in the latter case the lateral ventricles are continued back into the encephalocele. It is very necessary, when examining any tumour situated in the mesian line of the scalp, both in the child and the adult, to remember the existence of this class of swellings. In the child there is a direct communication between the tumour and the interior of the cranium, generally through the occipital bone immediately beneath the posterior fontanelle, and it is impossible to say at any period of life that the passage has closed. The skin covering the protrusion is sometimes remarkably firm, like the rest of the scalp; more frequently it is thin and flaccid. An encephalocele seldom gives rise to any symptoms unless by accident the skin be injured, when I have seen the fluid contents escape. In whatever way this escape of the contents is brought about, convulsions and death are the usual result, the case before mentioned being a singular exception. The only treatment that is admissible, is equable pressure to support the weakened skin, and even that will sometimes give rise to symptoms of cerebral disturbance, in which case it must of course be at once discontinued. Death usually happens at an early period, but if not, the communication with the brain diminishes as the ossification of the skull advances. The aperture may thus be reduced to a very small size, but the least communication suffices to render removal quite inadmissible. The tumour itself may also somewhat diminish with advancing years.

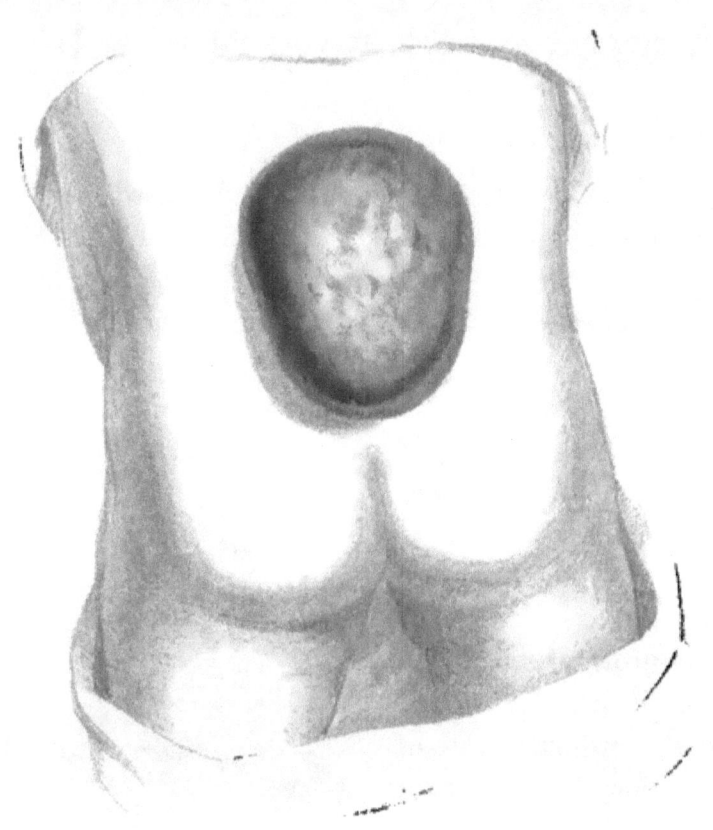

II. SPINA BIFIDA.

This is a meningocele of the spinal cord. It results from the deficiency of the laminæ of one or more of the vertebræ, and may occur in any portion of the column, though most commonly at the lower part. The accompanying lithograph represents the ordinary appearance of spina bifida, and the woodcut exhibits a section of the parts. The preparation from which it was taken is a fully grown fœtal spine, the lower part of which is in the ordinary state of spina bifida. The nature of this affection, and the state of the parts affected are well displayed. All the vertebral bodies were perfect, as also were the nerves. Four of the lumbar vertebræ have been removed. The arches of the three inferior lumbar vertebræ, 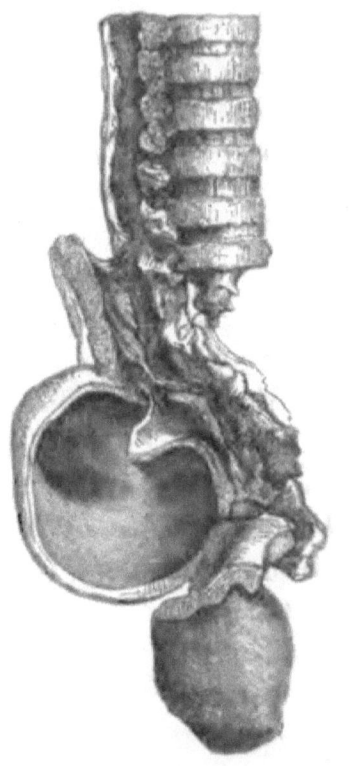 and all or nearly all the analogous parts of the sacrum, were extensively deficient, and only supplied by ligament at the third lumbar vertebra and at the lowermost part of the sacrum. A sac, the size of a large walnut, which was loosely filled with fluid, protruded.

Its covering was cuticular, but very delicate, and was at first highly vascular, as is shown in the lithograph. The lining was a loose, soft, tumid serous membrane, and had a wide communication with that of the medulla. The intermediate layer was a thick, firm cellular membrane, rather loose in texture, and fibrous. The most posterior of the sacral nerve fibres were seen coursing down the anterior wall of the sac, or the opening by which it communicated with the medullary serous canal.

The swelling in spina bifida is an extension of the membranes of the spinal cord, which lie immediately beneath the skin. Sometimes the termination of the cord in the cauda equina is spread over its inner surface. The sac is filled with fluid, which is a portion of the general cerebro-spinal fluid. The skin covering it is sometimes extremely thin and delicate, sometimes very vascular, and almost transparent, and appears to threaten every moment to give way. Unlike encephalocele, the base is generally broad. Upon manipulating the tumour, it feels soft and fluctuating, as the child lies upon its belly, in which position it is generally examined; when the child is erect the cerebro-spinal fluid gravitates into it, and it becomes more distended. Upon pressing the fingers around the edge of the tumour, the bony margin of the aperture is felt. For the most part the termination of these cases is in death; either suddenly, by giving way of the skin, or by a gradual failure of the strength. But the patient sometimes survives, especially when the meningocele is situated low in the back; and in these cases there may be seen in after life a firm dense swelling in the mesian line of the spinal column. These tumours are better not touched by the surgeon. Sometimes they do not become hard, nor shrink, but growing with the patient's growth they give rise to some little inconvenience.

Many plans have been recommended for the treatment of this affection. One of the most reasonable perhaps is that of repeated slight tappings. If the swelling be very tense, and give rise to much inconvenience, Velpeau has recommended injections with tincture of iodine; but I cannot understand the propriety of bringing an irritating fluid into contact with the delicate membranes covering the spinal cord. Operative proceedings for the removal of the tumour, though they have been variously employed, are absolutely forbidden by the anatomical relations of the parts involved. Giving an extra cuticular covering, by means of collodion painted once or twice daily over the tumour, is at any rate innocuous, and is likely to be beneficial; but the cases for which it is appropriate are generally very hopeless ones. The best treatment is mechanical support, by means of elastic bandages, pads of lint, air-pads, or any other contrivance that may seem most suitable Whatever relief can be afforded at all may be thus most easily and effectually given.

The woodcut in the following page shows a spina bifida situated in the region of the neck. The patient, aged three months, was admitted into Dorcas ward, under Mr. Hilton's care, in July, 1856. The sac was thin, and did not appear to contain anything but fluid. The child was in good health with the exception of having occasional attacks of laryngismus stridulus. No treatment was adopted, and it appeared very pro-

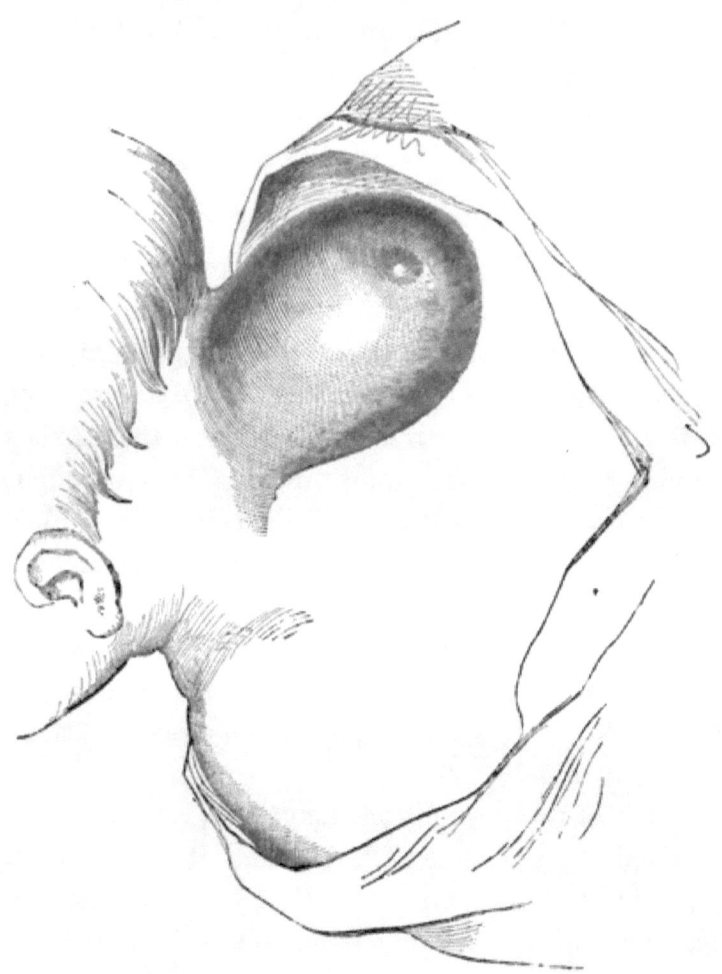

bable that this would become one of the cases of firm tumours situated in the mesian line of the back, to which I have before referred.

III. OCCLUSION OF THE MOUTH.

I have met with a case in which the mouth, at birth, presented an opening merely large enough to admit a full-sized bougie. I enlarged the aperture on either side, but unsuccessfully. The contraction returned to an even greater extent, and the child died. In a still-born but mature infant, I have witnessed occlusion of the eyelids and nostrils.

IV. IMPERFORATE RECTUM.

Of this malformation there are several varieties. First, the intestinal canal may terminate in the sigmoid flexure, the rectum being entirely absent; or, secondly, the rectum may terminate opposite some part of the sacrum. In both these cases, there exists a more or less perfect anus, opening into a pouch, between the termination of which and the end of the bowel, there intervenes a quantity of condensed cellular tissue. Thirdly, the rectum may extend as far as the anus, without any trace of that opening existing; and lastly, the rectum may terminate in the bladder, urethra, or vagina; in which case, also, there is not any anus to be seen.

Either of these conditions may exist, without being observed at the time of birth, and it is not until some twenty-four hours after, that the nurse notices that the child has not passed a motion. When the anus is absent, of course the nature of the case is soon discovered, and the surgeon's attention is called to it, so that relief can sometimes be given before distressing constitutional disturbance arises.

In this class of cases, in which the anus is absent, the surgeon's first object will be to ascertain whether or not the rectum communicates with any other channel. If this be the case, it is usually soon decided by the urine being stained with meconium, or an offensive mass accumulating in the vagina. If there be a communication with the bladder or urethra, operative interference may be attempted, though there is a very poor prospect of success. Still, cases are on record in which a trocar and canula have been pushed in the direction in which the rectum ought to be, and an opening has been established, with a permanently favourable result.

If the communication be with the vagina, there is much more encouragement for the operator. By introducing the finger into that passage, a tolerably good guide to the position of the bowel can be obtained; and either a sharp-pointed bistoury, or a trocar and canula, may be passed into it between the vagina and the sacrum. When a canal is thus established, however, there still remains the fistulous opening to be remedied.

When there is no anal aperture, but the rectum terminates low down, the probability of a successful issue to the operation is considerable. These cases are indicated by a bulging in the position of the anus, arising from the pressure of the distended bowel. Often a small central depression marks out, with great exactness, the spot at which the orifice should be. A crucial incision made here will give exit to the meconium. It has been recommended very strongly by Mr. Curling, that, if possible, the end of the rectum should be seized, dragged down to the margin of the anus, and stitched there. I have not, as yet, had an opportunity of trying this plan.

In all these patients, after the bowel has been opened, the utmost attention is necessary to many details, of which the most important is the passage, once or twice a day, of a bougie, gradually increased in size, for many weeks or months. Should the child live, it may be necessary to have recourse to the same treatment occasionally in after years. These are the class of cases in which attention is at once directed to the true nature of the imperfection.

The other class, in which the anus appears natural, are generally subject to domestic treatment, by purgatives, &c., before the surgeon's observation is specially drawn to the child. Accordingly, severe constitutional symptoms are present, the first of which is vomiting, the child appearing very uneasy and restless. The skin quickly begins to acquire a yellowish tinge, and the abdomen is distended; the child becomes drowsy, and the vomiting and distress continuing to increase, it soon dies, unless relief is afforded by operation.

The method to be pursued is always the same, it being impossible to tell at what point of its course the bowel terminates. These cases look very promising, but in consequence of the varying extent to which the bowel is deficient, it is impossible to calculate on the result of the operation. A trocar and canula should first be carefully passed through the anus and the pouch into which it opens, and carried in the direction of the rectum for the extent of an inch and a half or two inches. Care must be taken not to puncture the bladder in front, or the sacrum behind.

The following case, for which I am indebted to my friend Dr. Hicks, is an illustration of the most fortunate issue of this operation. The infant was born plump and strong, but after two or three days, the nurse noticed that no evacuations were passed from the bowels. Castor oil was given three times without result. After the third day, vomiting began, and Dr. Hicks's attention was called to the child. He examined the rectum with his finger, as far as it could be introduced, without finding any obstruction. On the fourth day, the vomiting of fœcal matter and blood was incessant, and the child's state was most precarious. The rectum was explored with a bivalve speculum, and Dr. Hicks was able to see a septum stretching across it, about two inches from the orifice. This septum was tense and bulging outwards. It was divided by a bistoury, and a large jet of meconium expelled from the opening. Shortly afterwards, the vomiting ceased, and the child became well in two or three days. The aperture was kept open by bougies passed occasionally for a period of two or three months. The child, five years afterwards, appeared in perfect health.

A result of the opposite kind is exhibited by a case on which I operated in February of this year. A female child, thirty-six hours old, was brought to me by Dr. Ladd; a squalid, emaciated object, already jaundiced, and vomiting frequently. The anus was perfect, and admitted a bougie for a distance of half an inch. I passed a trocar and canula in the direction of the rectum, and a quantity of meconium escaped. I had, however, from the appearance of the child, but little hope of its ultimate recovery. More meconium passed the following day, but the child sank. Upon post-mortem examination, the opening made in the bowel by the trocar was distinct, but there was also an opening into the vagina, and the rectum inclined towards that passage. There was at least an inch distance between it and the bottom of the anal pouch.

The following is a case of imperforate anus, in which the rectum terminated in the urethra. The patient was a child aged thirty-six hours, admitted into Guy's Hospital March 24th, 1856. A woman brought the child to the hospital and left it, saying only that it had never sucked, and had vomited. No further history could, therefore, be obtained. It was a well-grown, fat boy. The abdomen was much distended; the anus was found to be absent. The child was admitted about noon, and Mr. Cock immediately operated. No particular distension could be felt in the usual locality of the anus, but Mr. Cock passed a bistoury in the direction of the natural position of the rectum, and succeeded in at once opening the gut. Meconium followed in small quantity. It was afterwards perceived that the same matter flowed from the urethra. The child died at 8 p.m. Upon the post-mortem examination, no other deformity was found except the imperforate anus. The bladder and urethra were perfectly formed; the rectum also was perfect till it reached the position of the neck of the bladder, when it contracted into a very small canal, which immediately opened into the commencement of the urethra. The space between the termination of the rectum and the position of the anus, was about three-quarters of an inch in length. The incision opened the posterior surface of the rectum. The large intestine was enormously distended. The sigmoid flexure lay coiled over to the right side; and the intestines were still filled with meconium.

Artificial Anus.—If the bowel cannot be reached from the anus, the only other step is to form an artificial anus. There are two methods of attempting this operation: one known as Amussat's, which consists in making an opening on the outer border of the quadratus lumborum muscle; the other, called Littré's, which consists in opening the lower end of the colon in the left groin. Of these, in a child, the latter is to be preferred, although in performing the operation the peritonæum must be wounded, which it is not in the former case. This, however, in a healthy child, is not so serious a matter as might be supposed. An incision should be made about two inches and a half in length, in a direction level with Poupart's ligament, and about an inch above it. This incision must be carefully carried through the muscles until the peritoneum is exposed, when the distended intestine may be seen beneath it. The peritoneum being then carefully divided, the intestine is to be seized and stitched to the margin of the wound, to the extent of the opening to be made in it. When this has been done, the intestine should be opened by the knife, and the operation is complete. The careful application of a pad is necessary to prevent the intestine from continually discharging its contents, and great attention to cleanliness is demanded to prevent excoriation. Should the child live, several contrivances suitable to the case are available for the surgeon. The opening will gradually assume somewhat of the character of a sphincter.

V. CONGENITAL FRACTURES AND DISLOCATIONS.

It is reported that from falls or other accidents occurring to the mother, the fœtus in utero may have some of its limbs fractured or displaced. It has never been my fortune to see any such case; but several of the French surgeons give a lengthened account of them. They are said to have been very frequent in towns that had been recently bombarded; in which case they may perhaps be referred to strong and irregular uterine contractions induced by fear, or to the bad nourishment of the mother. Unless these cases are seen at the time of birth, congenital fracture is of course very difficult to diagnose, and congenital dislocation scarcely less so. In respect to *treatment*, it might appear reasonable, if a fractured bone (whether congenital or otherwise) be united in a bent or otherwise disadvantageous position, that it should be broken again, in order to procure a better union. For congenital dislocations no treatment, of course, is possible. A boy of three years old was brought to me with a congenital condition of the knee-joint, in which dislocation was producible at will. The tibia could, without difficulty or pain, be so far displaced from the femur, externally, that the finger might be distinctly laid on the outer half of its superior surface. The bone was easily replaced, but the child could not stand. He was also an idiot.

VI. EXCESS OR DEFICIENCY OF EXTREMITIES.

Deficiency of fingers or toes is not uncommon. Sometimes one or more extremity is altogether wanting. The following case of absence of the forearm was sent me in December, 1849, by my friend Mr. Robert Brown, of Brixton. I saw the infant at the age of three months. The parents were quite healthy and well formed; and this was their third child. There was no deformity known in either family. The deficiency was confined to one arm, in which the humerus was entire and of the normal length, but all the parts below were wanting. The skin at the end of the humerus was puckered. But of all deformities of this class, I have met with none more extraordinary than that exhibited by Dr. Hare at the Pathological Society, and recorded in volume x. of its *Transactions*. The subject of this malformation was a boy five and a half years old. He was almost entirely destitute of the upper, and absolutely so of the lower extremities. The head and trunk were natural. There was no deformity in the father's or mother's family.

In the case of supernumerary fingers or toes, which is still more frequent than their deficiency, the additional member is generally placed on the border of the hand, either the inner or the outer. The appendages are small, and appear half developed; the muscles, also, are wanting, and they are quite useless. Hence it is advisable to remove them. But it is necessary to be careful in doing so, as sometimes the joint by which they are attached is continuous with that of the finger from which they grow. If this should seem to be the case, it will be advisable to cut off the bone just above the joint; otherwise the removal may be effected at the articulation.

VII. WEBBED FINGERS.

The fingers, some or all of them, are occasionally connected by membrane, which may be of any extent, from a slight increase in the natural fold of skin to a complete union. This affection is frequently symmetrical, occurring between the same fingers on each hand. I have seen the middle and ring fingers so closely united that the former was prevented from obtaining its proper length, though the joints were bent, indicating the normal growth. Sometimes the joints between the second and third phalanges of adjacent fingers become united together, forming but one cavity.

As the child grows, it is remarkable how useful these fingers become without any surgical interference. And if this be attempted, the difficulty of keeping the fingers separate after division of the membrane is so great, that when cicatrization takes place, it is generally found that one quarter at least has reunited. If we take into consideration, also, the risk of mischief at the time of the operation, the length of time the wound takes to heal, and the very questionable utility of the parts afterwards, it is extremely doubtful whether, upon the whole, it would not sometimes be as well to leave them entirely alone. If, however, an operation be performed, it is generally advisable to pass the knife carefully between the fingers, piercing at the base, and carrying it forwards. Each finger is then to be enveloped in a piece of dry lint. As soon as suppuration is established, a loop of whipcord, or a thread of stout silk, is to be placed in the angle between the fingers, and fastened around the wrist; the wounds on the fingers being allowed to heal by granulation. After a time, however, despite of all endeavours, union between the adjacent surfaces occurs to a greater or less extent. This varies, however, considerably, according to the attention that is paid in the progress of the cure.

Several ingenious plans have been devised to attain a better result; such as transplanting a portion of skin, or making an incision at the base, and getting it to heal before the rest of the membrane is divided. But none have, as yet, a really satisfactory result.

VIII. CLUB FOOT.

Deformities about the feet are very frequent in children, but it is not my intention to treat of them in detail. Being seen, however, by every surgeon, and their treatment being comparatively simple, I should not be doing justice to my readers if I did not refer to them. The most common varieties of club-foot are the talipes equinus, varus, and valgus. But in the latter two affections, the inclination of the foot inwards or outwards almost always coexists with some degree of the raising of the heel, which constitutes talipes equinus. If these cases come early under the surgeon's notice, it is astonishing how much may be done without using the knife. At the same time, division of the tendons, not being a dangerous or difficult operation to any one conversant with anatomy, may be employed to expedite the cure. On this subject, I have nothing new to suggest. The most important part of the treatment consists in well-adapted and well-fitted mechanical appliances, which may require to be worn for months.

IX. HERMAPHRODITISM.

The varieties of this malformation are innumerable; in no case are any remedial measures available; and the only point of practical importance is to decide to which sex the patient belongs. In some cases, this is a matter of extreme difficulty, nay, even of impossibility; the enlarged clitoris and the small penis being so precisely similar, that unless the testicles have descended, no absolute distinction is possible.

The general characteristics, also, afford little help, the peculiar masculine or feminine qualities, always weak in children, being in these cases even less marked than usual. No special directions can be given for the diagnosis; it must be determined by a general consideration of the whole condition. And unless there be a decided preponderance towards one or other sex, the opinion must be held doubtful until it can be verified at a subsequent period, which may not be until puberty arrives.

X. DEFECT OF THE ABDOMINAL PARIETES.

It has been my lot, three or four times, to have cases brought under my notice in which the lower part of the abdominal parietes, just above the pubes, was wanting, the anterior wall of the bladder being also absent. Indeed, that viscus itself, in these cases, is reduced to a simple layer behind the anterior wall of the abdomen. There is no vesical cavity; the margins of the abdominal opening being continuous, or almost continuous, with the posterior wall of the bladder. The lining membrane of that organ is red, corrugated, and at its lower part may be seen the openings of the two ureters, distilling *guttatim* the urine, which, having no natural outlet by urethra, runs over the pubes, excoriating it as well as the upper part of the thighs. In the last case I saw, the child, though subject to this malformation, and consequent irritation from the urine, was otherwise well formed, stout, and hearty. It was the offspring of a Cornish fisherman. Operative surgery at present has done nothing for these patients; unless a case reported by Dr. Ayres, of New York, under the title of "Congenital Exstrophy of the Urinary Bladder," may be regarded as one of this class. In this case, there is said to have been a deficiency also of the pubes and anterior fourchette of the vulva. A plastic operation was performed, by first bringing down a portion of the abdominal integuments, and afterwards, by a second operation, closing up the lower portion of the opening. Attempts to collect the urine in an artificial receptacle are not attended with much success. The difficulty of adapting an apparatus to the abdominal walls, has not yet been overcome by the ingenuity of mechanicians.

CHAPTER XX.

TETANUS—SCALDS AND BURNS.

TETANUS.

TETANUS now and then occurs in children, though it is rare. The cases that I have seen have all been preceded by some injury, but this may be due to my position rather than to the rarity of the idiopathic form of the affection. Besides, children are so frequently subject to scratches or other injuries, that we are, perhaps, not justified in their case in assuming the certainty of a causative connexion between slight injuries and the convulsive disease.

A case which well illustrates the difficulty that may arise in reference to this question was admitted into Guy's Hospital this year. It is so admirably reported by Mr. F. P. Weaver, the then clinical clerk, that I cannot do better than give it in his words. It expresses all that I might otherwise say respecting the symptoms and course of the disease.

Case 1.—M. A. G., aged five years, residing at Walworth, was admitted into Guy's Hospital, Feb. 1st, 1860, under Dr. Barlow. She had some phthisical predisposition; the general health was good; had had measles and hooping-cough; the former three years, the latter a year and a half ago. She had continued well till a week before Christmas, since which time she has had a cough and become thinner, and has suffered from enlarged submaxillary glands. Three weeks before admission, she fell off a bed upon her head. Five days before admission, she was pushed down and fell on her face, her teeth cutting her upper lip, which has continued swollen since, and her right elbow being grazed. The next day, her mother observed her making wry faces.

The day before admission, she went to school as usual; came home at 4 P.M., quite excited, and jumped and danced about. Soon afterwards, when taking her tea, she suddenly threw herself back, complaining of pain, which was referred to the middle and lower part of the spine. What she had been drinking spirted from her mouth; her face became black and turgid; her hands clenched; her tongue was bitten; her jaws were fixed and rigid. In two or three minutes she became tranquil, but still complained of pain in her back and neck. Her face continued distorted. She passed a disturbed night, coughing frequently; the next morning she was brought to the hospital.

On admission, she was a strumous-looking child, of healthy complexion, rather emaciated; the upper lip was enlarged, as if strumous, owing, however, to the lip being inflamed from a slight wound; she had a small graze on the right elbow. The peculiar expression of countenance, called risus sardonicus, was evident; the brow was corrugated, the orbicularis palpebrarum rigid, the eyes were half closed, and the alæ of the nose and corners of the mouth drawn out. She could only partially open her jaw; the neck was stiff; pupils dilated equally; tongue moist and furred, with injection of the large papillæ; fauces slightly swollen and reddened; there was slight pain beneath the lower jaw, from enlargement of the left submaxillary gland. Her mind was undisturbed, and she could speak without difficulty. There was pain in the spine, about the seventh and eighth dorsal vertebræ, but no external appearance of injury nor distortion of spine. The pain was not increased by pressure. No loss of power or sensation in any limb. The hands usually assumed a fixed position; the wrists flexed, and the fingers rigid, extending straight out.

There was a slight cough; at times, prolonged fits of coughing, with some expectoration. The chest and heart sounds were normal; pulse 120, of fair strength. Abdominal organs normal; appetite bad; no vomiting; bowels rather confined; urine normal.

℞ Pulv. sodæ co., gr. viii. statim.
Ol. ricini, post horas tres.
Cataplasma lini gutturi.

Feb. 2.—In about the same condition; has taken some beef-tea without difficulty. Bowels have been relieved.

Pulv. Doveri, gr. iss.
Pulv. sodæ co., gr. v. quater die.

Feb. 3.—At nine last evening, was seized with rigidity of spinal muscles. This did not last for a minute, but returned about every twenty minutes until 11 A.M., when it ceased for an hour. It then returned every quarter of an hour during the night, and latterly has increased in frequency, coming on every three or five minutes. Has drunk beef-tea; bowels not moved.

11 A.M.—Lies on side, with head thrown back; face anxious; risus sardonicus less marked. Is conscious, and able to express her wants. Breathing easy; pulse 104. During a paroxysm, the head is thrown back, and there is a certain amount of opisthotonos; the heart beats rapidly; there is pain in the chest (diaphragm?), causing her to make pressure there; sometimes bites her tongue.

> Enema assafœtidæ, ʒviij. st.
> Cal. gr. j. tertiis horis, c. mist. sequente.
>
> ℞ Tinct. cinchonæ, ʒss.
> Vin. opii, ℳ iij.
> Sp. ammon. ar. ℳ v.
> Syr. aurantii, ℳ xv.
> Aquæ ad ʒj.

9½ P.M.—Paroxysms more frequent; are now brought on by touching her, besides being frequent when not touched. Bowels have not been moved.

> Enema haust. sennæ co. c. decoct. avenæ āā ʒiij.
> (Administered under the influence of chloroform.)
> Ung. hydrarg. vulneri admovend.

Feb. 4, 7½ A.M.—Paroxysms very frequent and severe; bowels freely moved; motions consist of large masses of solid fæces, not particularly fetid; has great pain during the attacks: is quite conscious; can take fluid into her mouth, but cannot swallow; cough troublesome. Chloroform administered; the room being kept quiet and dark. To have injections of beef-tea and milk.

10 a.m.—The paroxysms have been less frequent and less violent, with intervals of fifteen or twenty minutes; but they return on intermission of the chloroform; pulse 130, feeble; skin hot and dry. An ounce of sherry wine to be added to each injection.

>Cont. chloroform.
>Cont. enema c. tinct. opii, ♏x.

2 p.m.—Breathing much obstructed from mucus in the bronchial tubes. Chloroform at times intermitted, to allow the mucus to be coughed up; the paroxysms return when the chloroform is stopped.

7 p.m.—Breathing much obstructed. The child has not strength to cough up mucus. Pulse 150, very feeble; skin hot.

From this time she became rapidly exhausted, and died at 9 p.m.

Post-mortem examination, Feb. 6 (forty-two hours after death)—

Lungs posteriorly in a state bordering on pulmonary apoplexy, the parts being over-distended with dark blood; the large bronchial tubes reddened and containing much fluid; the small tubes not obstructed.

Spinal Cord.—Membranes natural, neither granules, adhesions, nor undue vascularity. Substance of medulla somewhat softened opposite the lower cervical vertebræ. No inflammatory changes were observed by Dr. Wilks on microscopical examination.

Brain unduly vascular (although the chest had

been first opened); on section, the cortical substance much darker than usual. In the white substance, "puncta vasculosa" very noticeable.

No other morbid condition was detected in the brain, except a small cyst in the pineal gland.

The following are concise accounts of two cases of undoubted traumatic tetanus:—

Case 2.—J. W., aged eleven, was admitted into Guy's Hospital, under Mr. Birkett's care, in March, 1854. Fourteen days before admission a cask fell on his foot, crushing his toe so as to necessitate amputation, and bruising his foot. The day previous to admission he complained of pain across his chest, and when raised in bed he became convulsed and rigid. His friends brought him to the hospital, with all the symptoms of well-marked tetanus. In two days the boy died. The treatment was chloroform, support, and wine. The post-mortem examination did not reveal any disease.

Case 3.—S. B., aged twelve and a half years, was admitted into Guy's Hospital on January 8, 1858, having injured his right hand by a piece of wood on December the 26th. On January 6th he came to the hospital among the outpatients, complaining of pain in the wound, which was suppurating, the hand was slightly swollen. He also complained of pain in his back, and felt very unwell. On his return home he became much worse, and on the following day tetanic symptoms supervened. On admission he did not appear very ill, although the symptoms were well marked. During the night severe opisthotonos made its appearance, and before daybreak on the following morning he died. At the post-mortem examination there was seen upon the ulnar side of the palm of the right hand a small hole, such as would have been inflicted by a blunt instrument. On slitting this up towards the wrist, a thin splinter of wood, half an inch long, was extracted. It was embedded in the substance of the ulnar nerve, the digital branches of which were exposed on the floor of the wound. No internal lesions were discovered.

SCALDS AND BURNS.

The distinction between a burn and a scald is not so simple as might be supposed. The injuries produced by boiling liquids are of very various degrees of severity. It is true that boiling water merely vesicates; but boiling oil, which has a much higher temperature, destroys the areolar tissue. It might be questioned, therefore, which name should be applied to such an accident. Practically, the distinction is of no moment, the principles of treatment being the same in all cases.

For my own part, I apply the name of scald to denote a simple vesication produced by heat, and that of burn to indicate destruction of tissue extending deeper than the epithelium, without reference to the mode in which the heat is applied. It is not my intention to enter at large into the appearances presented in these cases. Every elementary work upon surgery contains Dupuytren's classification of the six divisions of burns. But however theoretically perfect this may be, it is practically of value only as regards the prognosis and after consequences. My reason for referring to the subject is the frequency with which accidents from boiling water or from fire occur to children, though the treatment differs little from that which is applicable to adults. In consequence, however, of the delicacy of their organization, and their susceptibility to impressions, cases that might terminate favourably in an adult will often prove fatal in children. In this respect there is a striking contrast between their tolerance of burns and of other accidents. From contusions, wounds, operations, &c., children for the most part recover better than adults, but from burns they seem to recover much worse; the shock appears to prostrate them. It is true that an error is likely to arise here from our comparing the absolute and not the relative extent of surface injured, the same superficial area being, of course, much greater relatively to the child than to the adult; but even allowing for this, I think the fact is as I have stated. Nevertheless, it is undoubtedly the case, that children will recover from very serious and extensive burns: the deformities that are so frequent are sufficient evidence of this.

When the surgeon is first called to a case of burn, it is impossible to give a well-founded prognosis, for the reason that it is not possible to tell the depth to which the injury extends. Whether the child be burnt or scalded, there is at first the same uncertainty in this respect. A weak or delicate child, of course, has a worse chance than a strong one; and scalds or burns affecting the trunk are more unfavourable than those which are confined to the extremities. If the child survive the first shock, and do not die from the direct effects of the burn, there is still danger from secondary affections. Tetanus may follow, as in the following cases:—

Case 1.—M. W., aged nine years, was admitted into Guy's Hospital in November, 1846. His clothes were burnt on the upper part of the body, implicating half the right side, the right arm, and right leg. The cutis was exposed. The patient was sensible, though partially collapsed. Sloughing soon commenced, and the child became restless: ten days afterwards there was stiffness of the neck and difficulty of swallowing. Soon complete trismus set in; and he died two days afterwards in a spasmodic fit. The post-mortem examination did not exhibit anything particular.

Case 2.—E. H., aged nine years, was admitted into Guy's Hospital under Mr. Hilton's care, in April, 1855. The upper part of the body was burnt, through the clothes catching fire. The neck, upper part of the chest, and the arms were involved. She was not brought to the hospital for two days. The cutis was exposed. The girl was sensible, but much collapsed. For six days she progressed well, but then became very restless. On the morning of the seventh day the nurse found the child stiff, unable to swallow, and evidently suffering from tetanus. The little patient died during the afternoon, retaining her consciousness to the last. The post-mortem examination did not reveal anything to account for death.

In other cases, inflammation of vital organs follows, especially of the lungs, for which the intimate connexion that exists between these organs and the skin sufficiently accounts. It is said, also, that inflammation and ulceration of the duodenum are constantly present in fatal cases of burns; but the experience of **Guy's Hospital** does not bear out this statement.

In the *Treatment*, three distinct objects present themselves to the surgeon at different periods of the case: the first is the immediate relief of the external injury and shock; the second, to guard against the danger to life from subsequent complications; and the third, to prevent or remedy, as far as possible, the deformity which so often results when life is saved. In respect to the first of these, or the local treatment, the remedies that have been proposed are innumerable; and for the obvious reason that no particular application is of any special value. The grand principle of treatment in all cases of burns appears to be to exclude the air from the burnt surface. Undoubtedly the best means of doing this is to wrap the affected part or parts in cotton wool, and to apply a bandage. Whether the burn be superficial or deep, this should be the plan first adopted. The cotton wool should not be removed at all, unless the patient complain of uneasiness in the part, or there be profuse suppuration. But most frequently, by the end of the fourth day, in a severe burn, profuse suppuration has set in, and then the cotton wool must be removed, and other applications had recourse to. A favourite one with the nurses—and I believe it is the best—is common zinc ointment, varied occasionally with the resin ointment. Of course cleanliness is indispensable, but the dressings should not be changed more frequently than is absolutely necessary. I may throw out the suggestion whether it would not be greatly to the comfort of these little patients to be kept in a continual warm bath, until the healing process is well advanced.

A free use of some narcotic is necessary from the first, to relieve pain and secure some degree of sleep. The most nutritious diet should be given, and stimulants employed. Food, however, cannot generally be taken in large quantities.

If internal complications arise, depletive measures must not be had recourse to; the indication is rather for a more liberal measure of support. Should tetanus supervene, the rapidity with which it runs its course leaves little time for remedies.

The third object of the surgeon's care is to prevent deformity, if possible, or to cure it if it arise. In respect to the former of these objects, when the burn affects the face, and implicates the mouth, or nose, or eyelids, nothing can be done while the wound is healing. If the burn be on the neck or the extremities, much may be accomplished by the adoption of a proper position, and its maintenance by means of splints, &c.; but it should be remembered that these appliances must be continued for months after the wound has healed. In burns about the neck, Mr. Hilton is in the habit of placing these little patients at the end of the bed, with their heads hanging over it, so as to stretch the neck to its fullest extent, and no injurious effect on the cerebral circulation results; or a modification of the apparatus depicted at p. 118 will answer the same end. But the chief contraction occurs after cicatrization has taken place. When the extremities are the parts affected, much may be done by the application of splints, the limbs being kept as nearly straight as possible. By this means contraction may be rendered much less than it would otherwise be, but it is not to be entirely avoided. Every case has its own peculiarities, and the surgeon's ingenuity will be put to the test in devising means adapted to each.

With regard to operative proceedings to remedy deformity resulting from burns, it is my opinion that if all proper care have been taken to prevent deformity arising, when the burn is on the trunk or limbs, nothing further can be done. The tissues cannot be further stretched. But in any case in which due extension has not been practised during the healing, an operation may be of benefit. An incision should be made through the skin and areolar tissue, about one inch beyond the scar of the burn, and chiefly at the spots of greatest tension. The skin at once retracts, and leaves an exposed surface, which, in due course, will heal by granulation. At the time that this surface is cicatrizing, the greatest care is necessary to keep the parts well upon the stretch. Extension by splints, or otherwise, must be maintained, not only during the cicatrizing of the wound, but for months afterwards; and however long it is kept up, contraction, to a certain extent, sets in when it is discontinued.

If, by a burn affecting the neck, the face be distorted, a double operation may be had recourse to. The skin may be divided as described above, and in addition, incisions may be made in the face to free the constricted parts; the mouth, for example, or the angle of the eye. If the mouth be contracted to a very small aperture, we might be tempted to enlarge it by incisions on either side. But this should not be done. It would give no permanent relief, and would be very likely to aggravate the mischief.

In the treatment of these patients, much must be left to the surgeon's ingenuity. While these sheets were passing through the press, a little patient, on whom I operated at the Infirmary for Children two years and a half ago, came under my observation. In her case, the arm had been drawn to the side through a burn in the axilla, and the forearm also was flexed at an acute angle. I freed each of these parts, and kept the arm straight and away from the side, by means of a splint specially constructed. After six months' treatment, aided by the unremitting care of the excellent nurse, the girl went out with the surfaces completely healed. With the exception of not being able fully to extend the forearm, she has now the most perfect use of the upper extremity.

INDEX.

	PAGE
Abdomen :—	
Tapping of	110
Defect of abdominal parietes	326
Abscess :—	
Post-pharyngeal	76
Psoas	114
Anæsthetics	1
Anus :—	
Fistula in	94
Artificial	320
Antrum, tumour in	46
Artificial membrana tympani	304
Bladder :—	
Exstrophy of	326
Polypus of	160
Villous growth from	130
Bones, diseases of :—	
Caries	253
Rachitis	256
Burns and Scalds	334
Deformities from	338
Calculus in Bladder:—	
in female	132
in male	163
in urethra	139
Cancrum oris	23
Castration	190
Catheter, passage of	138
Cerumen	298

342 INDEX.

	PAGE
Cervical vertebræ, disease of	111
Chloroform	1
Cleft palate	39
Club-foot	324
Compression of brain	5
Concussion	5
Congenital deformities	308
Dislocations	328
Fractures	328
Phymosis	192
Contracted joints	282
Corns	295
Croup, operation in a case of	72
Cysts in neck	103
Deficiency or excess of extremities	322
Deformities, congenital	308
Discharges:—	
from meatus of ear	298
from vagina	123
Dislocations	252
Congenital	321
Dorsal vertebræ, disease of	113
Ear, diseases of:—	
Foreign bodies in meatus	296
General remarks on	306
Malignant disease of	306
Elbow joint:—	
Disease of	279
Resection of	281
Encephalocele	308
Epiphyses, separation of	250
Epispadias	191
Epistaxis	16
Extravasation of urine	145
Extremities, excess and deficiency of	322
Face:—	
Strumous ulceration of	21
Wounds of	11
Fingers, webbed	323

INDEX.

PAGE

Fistula :—
    in ano 94
    at umbilicus 107
Follicular growths in pharynx 76
Foreign bodies :—
    in larynx 55
    in meatus of ear 296
    in nose 13
    in pharynx 76
Fractures :—
    Congenital 321
    "Green stick" 250
    of head 5
    of nose 13
    Separation of epiphyses 250
Freezing mixtures 3

Generative organs, diseases of 122
Genu valgum 257
Glands, enlarged, in neck 99
    Tracheotomy performed for 71
Gonorrhœa :—
    in boys 195
    in girls 123
Growth from bladder 130
Growth from umbilicus 106
Gum, diseases of 40

Hæmatocele 187
Hare lip 29
Head, injuries of 5
Hermaphroditism 325
Hernia :—
    Inguinal 201
    Umbilical 199
Hip joint :—
    Disease of 263
    Treatment of 272
Hydrocele :—
    of cord 186
    of testis 183
Hypospadias 191

	PAGE
Imperforate rectum	315
Incontinence of urine :—	
in boys	155
in girls	135
Infantile leucorrhœa	125
Infantile syphilis	290
Injection of nævus (perchloride of iron)	233
Jaw, diseases of	41
Joints :—	
Contracted	282
Diseases of	259
Elbow	279
Hip	263
Knee	259
Wounds of	258
Knee joint, disease of	259
Larynx :—	
Accidents to	50
Foreign bodies in	55
Leucorrhœa infantilis	125
Ligature of nævus	237
Lithotomy :—	
in boys	168
in girls	132
Accidents in	178
Causes of death after	180
Lithotrity :—	
in boys	168
in girls	134
Lumbar vertebræ, disease of	114
Malignant disease of ear	306
Meatus of ear :—	
Discharge from	298
Polypus in	304
Membrana tympani, artificial	304
Moles	248

INDEX.

	PAGE
Mouth :—	
Congenital occlusion of	315
Diseases of	29
Nævus	206
Pathology of	210
Diagnosis of	212
Treatment	213
Classification of	215
by excision	218
by pressure	222
by ligature of vessels	225
by vaccination	225
by irritants and escharotics	226
by caustic introduced into substance	229
by seton	229
by heated wires and actual cautery	230
by needles and twisted sutures	231
by injections	232
with perchloride of iron	233
by ligature	237
by subcutaneous ligature	246
Nails, disease of	294
Neck :—	
Enlarged glands in	99
Operation of tracheotomy for	71
Cysts in	103
Noma	128
Nose :—	
Foreign bodies in	13
Fracture of	13
Polypus of	19
Thickened Schneiderian membrane	18
Nursing	4
Occlusion of mouth, congenital	315
Œsophagus, contraction of, from injury	81
Onanism	197
Operation for :—	
Castration	190
Fistula	96
Gastrotomy	83
Hare lip	32

	PAGE
Operation for:—	
Lithotomy:—	
in boys	168
in girls	132
Lithotrity:—	
in boys	168
in girls	134
Nævus:—	
Injection	233
Ligature	237
Polypus of ear	305
Polypus of nose	20
Resection of elbow	281
Sounding	166
Tonsils, removal of	46
Tracheotomy	66
Palate:—	
Cleft	39
Strumous disease of	42
Paracentesis abdominis	110
Paraphymosis	194
Penis:—	
Malformations of	191
Strumous ulceration of	196
Pharynx:—	
Contraction of—gastrotomy	81
Follicular growths of	76
Foreign bodies in	78
Phymosis	192
Piles	89
Polypus of:—	
Bladder	160
Meatus of ear	304
Nose	19
Rectum	93
Post-pharyngeal abscess	76
Prolapsus ani	89
Psoas abscess	114
Ranula	37

INDEX.

	PAGE
Rectum :—	
Fistula in ano	94
Imperforate	315
Piles	89
Polypus	93
Prolapsus	89
Ruptured urethra	136
Scalds and Burns	334
Deformities from	338
Schneiderian membrane	18
Skin, diseases of	286
Sounding for stone	166
Spina bifida	311
Spinal disease	111
Stone in bladder	163
Strumous disease of palate	42
„ ulceration of vagina	127
Syphilis :—	
Infantile	290
in boys	196
in girls	124
Testis, diseases of	188
Tetanus	328
Tongue-tie	36
Tonsils, disease of	43
Torticollis	104
Tracheotomy	66
Tympanum, disease of	298
Umbilicus :—	
Fœcal fistula at	107
Fissure of	105
Growth from	106
Ulceration of	105
Urethra :—	
Calculus in	139
Ruptured	136
Urinary Organs	122
Urine :—	
Extravasation of	145

	PAGE
Urine :—	
Incontinence of, in boys	155
in girls	135
Vagina :—	
Closure of	129
Discharge from	123
Strumous ulceration of	127
Warts	295
Webbed fingers	323
Wry neck	104

THE END.

www.ingramcontent.com/pod-product-compliance
Lightning Source LLC
Chambersburg PA
CBHW030359230426
43664CB00007BB/666